Jones

Guide to Short Fiber Reinforced Plastics

I0879688

SPE Books from Hanser Publishers

Roger F. Jones

Guide to Short Fiber Reinforced Plastics

With Contributions from
Mitchell R. Jones and Donald V. Rosato

Hanser Publishers, Munich

Hanser/Gardner Publications, Inc., Cincinnati

The Author:
Roger F. Jones, Franklin Polymers, Inc., Broomall, PA 19008-3413, USA

Distributed in the USA and in Canada by
Hanser/Gardner Publications, Inc.
6915 Valley Avenue, Cincinnati, Ohio 45244-3029, USA
Fax: (513) 527-8950
Phone: (513) 527-8977 or 1-800-950-8977
Internet: http://www.hansergardner.com

Distributed in all other countries by
Carl Hanser Verlag
Postfach 86 04 20, 81631 München, Germany
Fax: +49 (89) 98 12 64
Internet: http://www.hanser.de

The use of general descriptive names, trademarks, etc., in this publication, even if the former are not especially identified, is not to be taken as a sign that such names, as understood by the Trade Marks and Merchandise Marks Act, may accordingly be used freely by anyone.

While the advice and information in this book are believed to be true and accurate at the date of going to press, neither the authors nor the editors nor the publisher can accept any legal responsibility for any errors or omissions that may be made. The publisher makes no warranty, express or implied, with respect to the material contained herein.

Library of Congress Cataloging-in-Publication Data
Jones, Roger F.
  Guide to short fiber reinforced plastics / Roger F. Jones : with
  contributions from Mitchell R. Jones and Donald V. Rosato.
    p. cm.
  Includes bibliographical references and index.
  ISBN 1-56990-244-5. – ISBN 3-446-18479-1
  1. Fiber reinforced plastics. 2. Thermoplastic composites.
3. Thermosetting composites. 4. Injection molding of plastics.
I. Jones, Mitchell R. II. Rosato, Donald V. III. Title.
TA455.P55J66 1998
620.1′923—dc21   98-8480

Die Deutsche Bibliothek – CIP-Einheitsaufnahme
Jones, Roger F.
Guide to short fiber reinforced plastics / Roger F. Jones. With
contributions from Mitchell R. Jones and Donald V. Rosato. – Munich :
Hanser ; Cincinnati : Hanser/Gardner, 1998
  ISBN 3-446-18479-1

© Carl Hanser Verlag, Munich 1998
Production: Chernow Editorial Services, New York, NY
Printed and bound in Germany by Druckhaus Thomas Müntzer, Bad Langensalza

Foreword

The Society of Plastics Engineers is pleased to sponsor and endorse *Guide to Short Fiber Reinforced Plastics* by Roger F. Jones. This volume is extremely well organized and structured. The author's writing style has resulted in an easily readable presentation and an excellent reference for engineers and practitioners working in the area of fiber reinforced plastics.

SPE, through its Technical Volumes Committee, has long sponsored books on various aspects of plastics. Its involvement has ranged from identification of needed volumes and recruitment of authors to peer review and approval and publication of new books.

Technical competence pervades all SPE activities, not only in the publication of books, but also in other areas, such as sponsorship of technical conferences and educational programs. In addition the Society publishes periodicals, including *Plastics Engineering, Polymer Engineering and Science, The Journal of Injection Molding Technology, Journal of Vinyl & Additive Technology,* and *Polymer Composites,* as well as conference proceedings and other publications, all of which are subject to rigorous technical review procedures.

The resource of some 36,000 practicing plastics engineers, scientists, and technologists has made SPE the largest organization of its type worldwide. Further information is available from the Society at 14 Fairfield Drive, Brookfield, Connecticut 06804, USA.

Michael R. Cappelletti
Executive Director
Society of Plastics Engineers

Technical Volumes Committee:
Robert C. Portnoy, Chairperson
Exxon Chemical Company

Hoa Pham, Reviewer
The B.F. Goodrich Company

Preface

This book was written to provide a product design engineer with concise, basic information about the selection, use, and automated fabrication of short fiber reinforced plastics. These belong to an increasingly important class of materials that have been written about extensively from a theoretical standpoint, but not from a practical one. The authors have attempted to provide an examination of principal characteristics of short fiber reinforced plastics, so as to enable the design engineer to understand better what their advantages and limitations are in actual use, not a treatise on how to formulate or manufacture these composites. While not every material has been included, all commercially significant ones are represented, both thermoplastic- and thermoset-based. Throughout, the authors have striven to emphasize the practical aspect of the subject matter.

The typical design engineer is likely to have at least some general knowledge of plastics, and accordingly, introductory material is limited. Nevertheless, this is not intended to be an advanced or an exhaustive treatise, either, and the reader who is interested in theory or greater detail should refer to the references listed in the bibliography. These have been selected on the basis of usefulness as well as professional competence.

I would like to express my appreciation to the many people who made this book a reality. Among them are Ted Pilat, Mitch Jones, and Nick and Don Rosato for their conceptual contributions; Ashak Rawji, for the initial idea; Ed Immergut and Wolfgang Glenz for their support and motivation to finish; Michael Sepe, Robert Gallucci, and Seymour Newman for their critical reviews; Gabriele Eckler, Andrea Stoye, Elizabeth Hewitt, Elizabeth Gauger, Diane Actman, and Sonia Kennedy for their help with word processing and proofreading; personnel at BASF, Bayer, DuPont, Fiberite, General Electric, RTP, Instron, LNP Engineering Plastics, PPG, Vishay's Measurements Group, and other firms for contributions of data and figures; plus those many other friends and business colleagues, whom I cannot even begin to list, for their valued encouragement. Last, but not least, my wife and family for putting up with my long absences in the course of researching material for this work.

Roger F. Jones

Contents

1 Introduction

The economic application of plastics materials to mass-produced precision-engineered components has become possible largely as a result of the development of short fiber reinforced composites. Although these materials are far from new—the first thermoset-based materials appeared in the 1930s (long fiber reinforced thermosets had come out about 20 years earlier) and the first thermoplastic compounds in 1951—their optimum adaptation to mass production did not become a reality until the widespread use of the screw injection molding machine in the mid-1960s. Since then, the use of short fiber reinforced plastics has been growing rapidly and has now reached impressive commercial levels, estimated at over 530,000 metric tons in the United States during 1997, valued well in excess of one and a half billion dollars. Industry studies predict growth rates for these products will average 5 to 7% annually over the next 10 years, with thermoplastic-based materials growing three times as fast as thermosets.

Despite the length of time these materials have been available, new products continue to appear in the marketplace. Thus, short fiber reinforced plastics are still on the cutting edge of materials technology.

Just what are short fiber reinforced plastics? In the context of this book, these materials are defined, first, as composites of thermoset or thermoplastic matrices containing discrete fibers that do not exceed 10 to 15 mm or one half inch in length; they may also contain particulate fillers in combination with short fibers, for example, minerals, glass spheres, etc. The reason for choosing 10 to 15 mm as the cutoff point to distinguish between short and longer fiber reinforcements is that this is the prevalent break point between materials fabricated via automated mass production, such as injection molding or extrusion, rather than nonautomated or semiautomated processes such as hand layup, sprayup, rotomolding, compression and transfer molding, etc. Short fiber reinforced materials are the more commercially important materials and the ones that exhibit the most significant development and growth. This aspect thus also becomes a second definition of the materials described in this book: that they can be fabricated by automated equipment.

Some suppliers offer "long fiber" or "long glass" reinforced thermoplastic materials, but these products still fall within the above definition and accordingly are treated as variants of short fiber reinforced plastics. As defined by long fiber compound suppliers, their materials contain fiber rein-

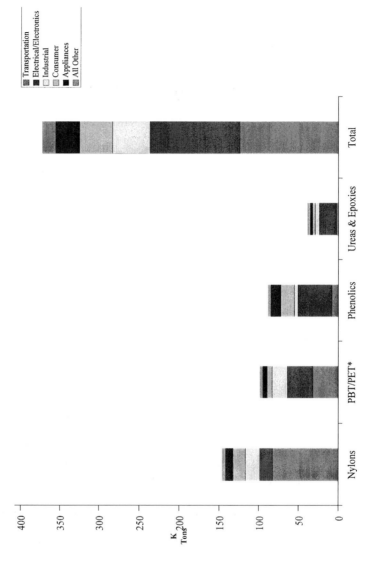

Figure 1.1 1995 end-use markets by major polymer

Table 1.1   Short Fiber Reinforced Plastics 1997 End-Use Markets by Major Base Polymers, thousands of metric tons

Category	Nylons	PBT/PET*	Phenolics	Ureas and Epoxies	Total	%
Transportation	93	37	7	2	139	34
Electrical/electronics	15	36	39	20	110	27
Industrial	25	21	5	5	56	14
Consumer	28	6	15	3	52	13
Appliances	10	6	12	3	31	7
All Other	12	4	4	3	23	5
Total	183	110	82	36	411	100

*Thermoplastic polyesters: PBT = polybutylene terephthlate, PET = polyethylene terephthlate
Source: Franklin Polymers, Inc.

forcements that average 10 to 15 mm in length, while "short fiber" compounds contain fiber reinforcements that are something less than this, but average perhaps 7 mm (0.25 in) in length. The starting length of the fibers is certainly of major importance, but the length of the fibers in the finished part is also of great relevance, as there is minimal property enhancement of the matrix resin if the reinforcement fibers fall below the minimum critical length, so that careful molding is vital to avoid excessive attrition of the fiber length. Nevertheless, when properly processed, long fiber materials can confer significant property improvements over materials that contain shorter fiber lengths, particularly in the areas of resistance to crack propagation and long-term creep.

The purpose of reinforcing plastics is to increase dimensional stability, strength, toughness, heat and environmental resistance, as well as the reproducibility and hence predictability of these properties. These enhanced qualities have led those in the industry to categorize reinforced plastics as "engineering plastics." In many end-uses, these features plus high strength-to-weight ratios, ease of fabrication, and the ability to design consolidated one-piece parts, make relatively high-cost reinforced plastics more economic in use than lower cost conventional materials such as metal, wood, or glass. Improved function, especially durability, is an important advantage over traditional materials for almost every plastic material.

In 1997, the principle end-use markets for the largest volume composite materials (representing slightly more than 70% of the overall total) were as shown in Table 1.1 and Figure 1.1. Over the past two decades, phenolics and other reinforced molding grade thermosets has grown about 25%, while that of nylon and other reinforced thermoplastics has grown over 300%. Currently, the volume of reinforced thermoplastics used is about double that of reinforced thermosets. Electrical/electronics and automotive uses account for the majority of all short fiber reinforced plastics applications.

2 Materials

As mentioned in Chapter 1, short fiber reinforced plastics are composites of reinforcing fibers in plastics matrices. This chapter attempts to characterize the principal component materials, to provide a better understanding of the balance of properties that they each contribute to the resulting composite.

First, the principal types of reinforcements and additives are reviewed, then the thermoplastic and thermosetting resins that constitute the matrices.

2.1 Reinforcements

Reinforcements primarily give strength and stiffness to the composite. The predominant fibers used for reinforcement are made of glass or carbon (graphite). Polymeric and metal fibers have use in specialized circumstances. Mineral fibers usage is low and declining because of health concerns and lower performance, although these materials were once of considerable commercial importance. Natural fibers have yet to demonstrate performance levels to justify their usage in any but nondemanding applications. Certain fillers used in conjunction with fibers, to enhance specific properties, are also described.

2.1.1 Glass Fibers

Glass fibers are the most widely used form of reinforcement for plastic materials. Advantages of glass fibers over other reinforcements include a favorable cost/performance ratio with respect to dimensional stability, corrosion resistance, heat resistance, and ease of processing.

Glass fiber reinforcements are produced by drawing filaments of glass from a furnace containing molten glass, combining them into bundles, applying surface treatment, and chopping into specific lengths. The surface treatment, which consists of lubricants, sizings, and coupling agents, is necessary to reduce strand breakage during processing and provide compatibility with different resin systems.

Most short glass reinforcements are made from E-glass, a calcium–aluminum–borosilicate formulation originally developed for electrical applications; its good balance of properties and low cost has broadened its uses to virtually all applications. In addition to excellent dielectric and mechanical characteristics, E-glass has good resistance to heat and water, fair resistance to bases, and low resistance to acids. A variation in formula, E-CR-glass, offers improved resistance to acids, as well as bases and water. The principal physical properties are shown in Table 2.1. S- and R-glass, high-strength reinforcements (some 30% greater than "E" glass) are typically offered in continuous roving but not in chopped form, nor with a broad range of thermoplastic coupling agents. S- and R-glass fibers also cost significantly more than "E" glass—up to seven times more. The other principal type of glass is "A" glass, a soda–lime–silica composition more typically used for windowpanes and bottles.

Although chopped fiber is available in lengths 3.2 mm (0.125 in.) on up to 25 mm (1 in.) some applications require the use of milled fiber. Milled fiber, as its name implies, is chopped fiber that has been ground in an attrition mill to shorter lengths than can be obtained by chopping—typically 1.6 to 0.8 mm (0.0625 to 0.03125 in.). Milled fiber is used in applications where stiffness, dimensional stability, high flow during molding, and more uniformly isotropic mold shrinkage are considered more important than tensile strength or toughness.

Although glass fibers can be made in a variety of diameters, those used in short fiber reinforced plastics are typically 13 μm (0.0005 in.) usually referred to by the letter designation "K."

The theoretical minimum critical length for a reinforcing fiber has been estimated variously to be about 50 to 100 times the diameter. However, several investigators have shown that 95% of the fiber strength can be developed in a composite in which the length-to-diameter ratio (L/D) is as low as 10:1 [1 to 3]. For "E" glass, this would be only 130 μm (0.005 in.). Nevertheless, if we want to achieve 99.5% of the strength of an "E" glass fiber rather than only 95%, then the L/D ratio must be 10 times this value, or 1300 μm (0.05 in.). Since fiber lengths in typical molded parts show a range of 1.5 to 2.5 mm (0.06 to 0.01 in.), it is clear that if care is taken to avoid above-normal fiber length attrition in tool design and molding, the properties obtained in a part should correspond to data sheet values. Of course, this requires that the use of reworked scrap be minimized (seldom is more than 25% desirable). It is also necessary to compare data sheet values for the material to be used with those from other suppliers to ensure that the start-

Table 2.1 Typical Physical Properties of Representative Fibrous Reinforcements

Property	E-Glass	Carbon (PAN)	Aramid	PET	Chrysotile asbestos	Sisal	Stainless steel	Ceramic
Fiber diameter (mm) (in.)	0.0102 (0.0004)	0.0076 (0.0003)	0.0127 (0.0005)	0.0229 (0.0009)	0.0051 (0.0002)	0.2540 (0.0100)	0.0076 (0.0003)	0.0051 (0.0002)
Specific gravity	2.54	1.84	1.45	1.38	2.50	1.50	7.77	2.7
Modulus of elasticity (MPa) (Mpsi)	72.4 (10.5)	359 (52)	131 (19)	10.0 (1.4)	159 (23)	16.5 (2.4)	193 (28)	103 (15)
Tensile strength (GPa) (Kpsi)	3.45 (500)	3.79 (550)	2.76 (400)	1.03 (150)	2.07 (300)	0.52 (75)	0.59 (85)	1.72 (250)
Tensile elongation (%)	4.8	1.1	2.4	22	NA	2–3	2.3	NA
Thermal conductivity (W/m °K) (BTU-in./hr-ft²°F)	1.01 (7.0)	8.65 (60)	0.50 (3.5)	0.25 (1.7)	NA	NA	1.12 (7.8)	2.88 (20)
Coefficient of linear thermal expansion (C×10⁴)	0.08	0.06	NA	NA	NA	NA	0.04	NA
Approximate relative cost (by weight)	1	15	4	2	1	0.7	18	3

ing material has been properly formulated so as to offer adequate reinforcement.

Typical physical properties of E-glass fiber are shown in Table 2.1. Costs of fibers are stated as ratios where glass fiber has a value of 1.0; when this book was written, the actual cost of glass fiber was about $1.87/kg ($0.85/lb).

2.1.2 Carbon Fibers

Carbon fibers are truly a "space-age" development. The dominant end-use for carbon fiber reinforced materials has indeed been aerospace, where cost is typically secondary to performance. Although there are a wide variety of types of carbon fiber produced for use in composites, varying, for example, in degree of graphitization and diameter, there are only a limited number made for short fiber reinforced plastics. Two principal types are available in chopped and sized form, polyacrylonitrile (PAN)-based and pitch-based. Differences in mechanical properties between the two forms in a composite greatly favor PAN over pitch, as shown in Table 2.5. Therefore, the focus of this work is on PAN type carbon fiber as the far more commonly used; typical physical properties are shown in Table 2.1. Pitch-based carbon fiber is less expensive than PAN-based, and it finds some use in applications in which the primary requirement in the composite is electrical conductivity rather than ultimate mechanical properties.

Carbon (or graphite) fibers are produced by pyrolysis, a high-temperature reduction of the precursor polymer to its basic continuous carbon atom chain. Compared to glass fibers, carbon fibers offer higher strength and modulus, lower density, outstanding thermal and electrical conductivity, but much higher cost. In addition, they are very chemically resistant and naturally slippery. The low density and high mechanical properties of carbon fibers allow great flexibility in formulating composites with specific performance characteristics, including using mixtures of glass and carbon fibers (to reduce the cost of using carbon fibers alone). The high electrical conductivity of carbon fibers is also ideally suited for electromagnetic shielding applications. The excellent thermal conductivity, as shown in Table 2.1, helps extend the fatigue resistance of composites where failure can be caused by internal heat buildup.

Since carbon fibers are normally manufactured in thinner diameters than glass fibers, the minimum critical length is less, providing a greater safety margin for property enhancement.

2.1.3   Organic Polymeric Fibers

Organic polymeric fibers include aramid (aromatic polyamide) and poly-ethylene terephthalate (PET) fibers. Typical physical properties of DuPont's Kevlar® aramid and Allied Signal's Compet® PET fibers are shown in Table 2.1.

Aramid fiber is available in chopped and sized form. Composites made with aramid fiber are useful in applications in which abrasion resistance and low wear on mating services are critical. Because the fibers show low compressive strength, they are easily damaged (shortened) during compounding; hence, the mechanical properties of aramid composites are not particularly noteworthy (see Table 2.5), especially in view of the cost. Aramid fibers exhibit a curled or "kinked" appearance unlike the straight rodlike appearance of glass or carbon fibers. This characteristic tends to make aramid-containing composites resistant to alignment in the direction of flow during processing and hence more uniform in property distribution.

PET chopped strand is used to blend with glass fiber to upgrade impact strength in brittle resin matrices. PET also does not confer exceptional mechanical strength or stiffness, but offers lower cost relative to other non-glass reinforcements. As a side benefit, PET fiber is much less abrasive on mold surfaces than glass. It is typically used in thermosets and is useful up to about 204 °C (400 °F) service temperature (of the composite), but it will begin to melt at 246 °C (475 °F).

2.1.4   Mineral Fibers

Mineral fibers were once very popular, but have been rarely used since about 1975. Asbestos, although an excellent high-strength, low-cost reinforcement, has been the subject of much legislation, regulation, and litigation in view of the potentially adverse health effects resulting from excess inhalation of free fibers. As a result, material suppliers and fabricators no longer utilize asbestos in composites. Use of some "whisker" type minerals has faced the same basic health questions and suffered the same fate. Acicular or needlelike minerals, although offering some improvement in the physical properties of a composite, vs. granular or spherical shaped fillers, generally do not have a sufficiently great L/D ratio to reinforce effectively and, therefore, seldom compare favorably with fibers on a cost–performance basis. Wollastonite, calcium metasilicate, is an acicular mineral resembling a fiber, but its L/D ratios run as low as 3:1. Wollastonite can enhance the ability of the composite to be electroplated, and may give better surface appearance in the finished part than fiber reinforced composites.

2.1.5 Natural Fibers

Natural fibers are used almost exclusively in low-severity applications for thermoset composites, and include primarily α-cellulose (alkali-treated wood pulp), cotton, sisal, and jute fibers. Although they provide some improvement in stiffness and impact resistance, their use is severely limited by their relatively low heat resistance: strength loss sets in around 124 °C (255 °F) and thermal degradation commences around 163 °C (325 °F). Also, they impart a dark coloration to the composite, tend to degrade quickly on exposure to sunlight and microbial attack, and to absorb water and oils, with attendant diminished mechanical and dielectric properties. These fibers generally have low aspect ratios, which limits their ability to improve strength and toughness properties. Their principal benefit is low cost and, hence, they can be viewed more as fillers or extenders than as reinforcements. Typical properties of sisal fiber are shown in Table 2.1.

2.1.6 Metal Fibers

Metal fibers include stainless steel, aluminum, and nickel-plated glass or carbon fibers. These fibers are typically used in composites intended for applications requiring electrostatic charge dissipation or electromagnetic frequency shielding. They are not optimal for reinforcement because they tend to curl up during processing, but the low concentration of fiber (normally 5 to 10%) required to achieve satisfactory shielding performance usually does not degrade the mechanical properties of composites based on them to unacceptable levels. Still, toughness and modulus levels are often below those of conventional carbon or glass fiber composites. The principal advantage for metal fibers over carbon fibers for shielding is lower cost per unit reduction in surface resistivity. Stainless steel fibers are the most widely used at present.

2.1.7 Ceramic Fibers

Ceramic fibers (other than glass) include alumina, boron, silicon carbide, alumina–silica, and other metal oxide–silica fibers. The physical properties of these reinforcements in matrices compare favorably to glass and other fibers, particularly with respect to compressive strength and general retention of properties at elevated temperatures. The properties of an alumina–silica fiber, the most widely used, are shown in Table 2.1. Ceramic fibers suffer from two significant limitations: cost (from about $4.4/kg ($2/lb) up to $1100/kg ($500/lb)) and an inherent brittleness that leads to appreciable fiber length attrition during compounding. Also, alumina fibers closely resemble asbestos. They are used primarily in fluoropolymers and thermoset resins in aerospace, chemical process equipment components, and brake lining applications, but alumina fibers are being phased out because of the asbestosis potential (as is the case with mineral fibers, noted previously).

2.1.8 Nonfibrous Fillers and Additives

Nonfibrous fillers and additives are sometimes used in conjunction with fibers to reduce cost and enhance certain specific properties, although this often comes at an offsetting disadvantage of reducing one or more other properties (see Table 2.2). For purposes of this discussion, mention has been omitted of basic additives that are routinely added in small quantities (less than 1% by weight) to unreinforced polymers, such as heat stabilizers, mold releases, and processing aids. The principal types of the subject materials include those described in the following.

2.1.8.1 Glass Spheres

Glass spheres or beads, when added to glass fiber reinforced composites, reduce anisotropic mold shrinkage and improve surface appearance. Some grades are available with surface coupling agents added. Virtually all glass spheres used for injection molding composites are solid. Hollow spheres generally do not have sufficient crush strength to withstand the pressures developed during injection molding. Solid glass spheres for use in plastics are offered in both A- and E-glass compositions. Although diameters ranging from 5 to 5000 μm are commercially available, the size typically used for plastics is 30 μm.

Depending on the loading level and presence (or absence) of coupling agents, strength properties and deflection temperature may be adversely affected—the same geometry that facilitates flow provides little for the resin matrix to "grip." As is typical of many fillers, addition of glass spheres will increase the stiffness of the base resin but lower the ultimate elongation to failure. Table 2.2 shows some comparative properties of nylon 6/6 containing glass fibers vs. glass spheres.

Table 2.2 Effect of Nonfibrous Fillers on Nylon 6/6 Composites

Glass fiber, by weight	40	25	0	0
Mineral filler (kaolin)	0	15	40	0
Glass spheres	0	0	0	40
Specific gravity	1.46	1.47	1.49	1.44
Tensile strength MPa	214	138	103	97
(Kpsi)	(31)	(20)	(15)	(14)
Flex modulus GPa	11.0	10.0	6.6	5.2
(Kpsi)	(1600)	(1450)	(950)	(750)
Izod impact, notched (6.4 mm)	139	43	43	53
J/m (ft-lb/in.^{3})	(2.6)	(0.8)	(0.8)	(1.0)
Relative cost by volume	1.00	0.98	0.85	0.92

Source: LNP Engineering Plastics, Inc.

2.1.8.2 Minerals

Minerals, such as kaolin (clay), talc, mica, silica, feldspar, barium sulfate, calcium sulfate, and calcium carbonate, are also used to reduce anisotropic mold shrinkage, improve stiffness, and lower cost. Mineral fillers as a rule offer good chemical and heat resistance, low moisture absorption and thermal conductivity, and good dielectric properties. Unfortunately, they also tend to lower mechanical strength properties of the reinforced composite and often impart variable coloration. Preferred minerals, such as kaolin, talc, mica, and calcium carbonate, usually have platelike rather than granular forms, so that their geometry (and hence reinforcing effect) will tend to resemble more that of fibers than spheres. The mean diameter of mineral fillers varies widely, from around 5 up to 500 μm. Many are offered with coupling agents added to improve adhesion to the resin matrix. Table 2.2 provides an insight as to the effectiveness of glass fiber vs. mineral filler, and a mixture of the two.

Feldspar is translucent in compounds, and is hence used in such applications as polypropylene battery cases. Barium sulfate is added to compounds where X-ray opacity or added density is desired.

2.1.8.3 Elastomers

Elastomers, particularly acrylic, polyolefin, polyester, and polyurethane types, are added to improve the matrix toughness, especially at low temperatures. Their use normally results in an improvement in both ultimate and yield elongation but at some sacrifice in strength, modulus, heat deflection temperature, and cost. Elastomers also usually confer some improvement in surface abrasion resistance. The efficiency of the modification depends greatly on the compatibility between the elastomer and the resin matrix, as well as the particle size and degree of dispersion. This is a complex area of technology in its own right, and the sophistication of the compounder who formulates and processes the blended composition is critical to the outcome.

2.1.8.4 Flame Retardants

Flame retardants enhance the composite's ability to either resist ignition or sustain combustion after having been set afire by a temporary source of ignition. This does not mean that the composite has been made "nonburn-ing"; all polymers are consumed in open fires. Usually, two or more compo-nents, a "system," are required to quench ignition and retard combustion. For example, antimony oxide and a chlorinated cycloaliphatic compound constituted one of the earliest flame retardant systems for polypropylene. Many systems require high loadings to be effective, in some cases as much as 40% by weight. This can make the resulting composites dense and brittle, and hence require reinforcement to be used in most applications. Systems that use halogenated components frequently have a narrow processing range because of limited high temperature stability; chemical decomposi-tion of the system also yields corrosive vapors. Although other systems may not have these particular defects, they may offer other problems, such as lower arc resistance or reduced mechanical properties; each system must be compared carefully. See Table 2.4 for illustrations of the effects on nylon 6/6 and PBT composite properties.

2.1.8.5 Lubricants

Lubricants, such as silicone oil, powdered polytetrafluoroethylene (PTFE), powdered high molecular weight polyethylene (HMWPE), molybdenum disulfide (MoS_2), and graphite powder, are all used to enhance the wear resistance and friction characteristics of composites. Ideally, the enhance-ment will last for the life of the component, and no additional lubrication, such as oil or grease, will be needed in service for the part to perform. Because lubricants are used in relatively small amounts, the effects on other mechanical properties are generally negligible, other than limiting color-ability: PTFE and HMWPE impart a creamy tint, whereas graphite and MoS_2 produce a black color; only silicone oil is colorless. Silicone oil may be unde-sirable for use in electronic applications, however, because it can migrate at high temperatures to electrical contact surfaces and interfere with the oper-ation of the component. Thorough dispersion of the lubricant in the com-pound is essential to its efficient performance in an application.

2.1.8.6 Coupling Agents

Coupling agents are the molecular "glue" that bond the reinforcement and the polymer matrix together in the composite, enhancing mechanical and other properties. Although coupling agents are considered highly proprietary by compounders, the vendors of these materials have published enough about them for us to know that organosilanes, titanates, zirconates, and acrylic and other organic acids are in use. Most reinforcements are supplied to compounders with the coupling agents already applied, ready for compounding. Some polymer systems, particularly acetal and polypropylene, require additional, specific coupling agents added during compounding to achieve more than minimal bonding of the polymer to the reinforcing fiber; these compounds are usually identified by suppliers as "chemically coupled."

2.1.8.7 Blowing Agents

Blowing agents are substances added to thermoplastics during processing to achieve a cellular structure. For composites, this may be desirable for several reasons: first, a low level of foaming may be useful to eliminate sink marks from the surface of a part where a thick section is showing greater shrinkage than the surrounding sections. Second, a greater level of foaming may be desirable to form a "structural foam." Structural foams are used for larger parts, to reduce part weight, cut fabrication cost, and enhance sound-deadening characteristics and thermal insulating properties.

Chemical blowing agents that are used for composites decompose at processing temperatures to release an inert gas, usually nitrogen. The principal blowing agents sold for these applications include azodicarbonamide, p-toluenesulfonyl semicarbazide, and 5-phenyltetrazole. Although these are relatively expensive materials, only a very small amount, 0.1 to 0.8%, is required. They may be dusted on the resin pellets, but the preferred method is to use concentrates (the blowing agent dispersed in a resin compatible with the composite) tumble-blended with the compound pellets or added via a weigh feeder.

An inert gas, such as nitrogen or carbon dioxide, may also be added directly to the polymer melt during processing to produce the foams described previously. The results are the same. There are patents on this technique, but licenses are readily available via the manufacturer of the molding machine to be used.

2.2   Plastics Resins

This section deals with the principal plastics matrices in which the reinforcements are dispersed. Plastics are the cement that hold the composite together and the medium that allows low cost (e.g., low temperature) fabrication of the composite. However, a plastics matrix also generally has lower mechanical, thermal, and other properties than the reinforcement, and thus often limits the performance characteristics of the composite. It is therefore important to recognize just what advantages and disadvantages each plastic material brings to a composite.

2.2.1   Thermoplastics

Thermoplastics are polymers that can be resoftened more than once—the molecular chains of these polymers do not crosslink and remain essentially unchanged (except for some shortening, usually minor) each time they are processed. This is in contrast to thermosets, which undergo a chemical reaction when processed, and hence cannot be resoftened. Thermoplastics usually cost less to process than thermosets in terms of reusable scrap and faster cycle time. Thermosets, however, generally have greater dimensional stability and heat and chemical resistance than thermoplastics. Both thermoplastics and thermosets can sometimes also be physically blended with each other, offering a broadened range of cost and performance.

Thermoplastics are also further categorized according to their morphology, which is generally either amorphous or semicrystalline. The principal difference between these two types may be observed as the temperature is raised while the material is under stress, as shown in Figure 2.1. Crystalline materials tend to show little change until the glass transition temperature (T_g) is reached, when a reduction of 50% or more in strength and stiffness is observed; this can be significantly offset by reinforcement, but either way, the composite will continue to have at least some load-bearing utility on up to the crystalline melt point. Amorphous materials, on the other hand, show a gradual diminishment of mechanical properties until their glass transition point is reached; reinforcement does almost nothing to improve their load-bearing properties above the T_g temperature. Crystalline polymers also generally have higher solvent resistance and load-bearing properties at high temperatures than do amorphous polymers. On the other hand, amorphous polymers often exhibit better dimensional stability, with less shrinkage and warpage after processing.

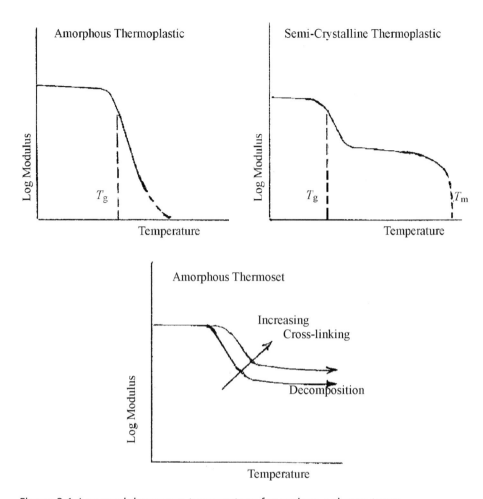

Figure 2.1 Log modulus versus temperature for various polymer types
(After S. Newman)

The listing of thermoplastics resins used for short fiber reinforced composites is arranged alphabetically. Major properties of the unreinforced polymers are shown in Table 2.3 and those of typical reinforced versions in Tables 2.4 and 2.5; these data represent a range of values from manufacturers' data sheets. Because material prices are subject to constant fluctuation, the cost relationships are expressed as ratios, where polypropylene has a value of 1.0; at the time this work was written, polypropylene cost $0.99/kg ($0.45/lb); even so, these cost comparisons should be regarded as only approximate. Resins shown are all commercially available and listed as the matrix of a composite in the product literature of a supplier. A notable exception to the polymers included in this chapter is acrylic: there are no commercially made compounds based on this material.

The data furnished here on materials and compounds are based on prime, virgin polymers. Although there is some demand for postconsumer recycled (PCR) polymers, their availability is limited and field performance results not yet fully documented. An important exception is nylon 6 and 66 compounds made from reprocessed textile scrap. There are a number of reputable compounders who have supplied these materials for over 25 years without problems; short-term physical properties are virtually identical to those of compounds based on prime polymers. Nevertheless, they have been exposed to at least two or three more heat histories than prime polymers and therefore may have slightly shortened service lives.

2.2.1.1 Acrylonitrile–Butadiene–Styrene
Acrylonitrile–butadiene–styrene (ABS), although nominally a terpolymer, is actually an amorphous blend of styrene–acrylonitrile (SAN) copolymer and SAN-grafted polybutadiene rubber particles. The ability to vary the nature and amount of these chemical building blocks permits the supplier to formulate a range of toughness, rigidity, surface appearance, and ease of processing. The moderate heat, moisture, and chemical resistances of ABS are its limiting properties. Glass fiber reinforcement at low loadings (up to 20%) enhances its dimensional stability; flame retardant (glass reinforced) grades are also available.

2.2.1.2 Acetal
Acetal or polyoxymethylene (POM) is available both in homopolymer and copolymer form. Both types are semicrystalline and offer high strength, stiffness, a low coefficient of friction, dimensional stability, and exceptional chemical and moisture resistance. Glass fiber reinforced grades show further enhancement of strength and stiffness; the copolymer composite has significantly better mechanical properties than the homopolymer composite, particularly if exposed to hot water for any length of time, and has a wider processing temperature range. Unfortunately, no flame retardant grade exists, although acetal's dielectric properties are quite good.

2.2.1.3 Fluoropolymers
Fluoropolymers are a large family of semicrystalline materials that have outstanding chemical inertness, high temperature resistance, and low friction. Dielectric properties are excellent and stable over a wide temperature and frequency range; all of these polymers are inherently flame retardant. They

Table 2.3 Representative Properties of Major Thermoplastic Resins (Unreinforced)

Type	Specific gravity	Melting or softening point °C (°F)	Tensile modulus GPa (Kpsi)	Deflection temperature °C @ 1.82 MPa (°F @ 264 psi)	Notched Izod impact strength J/m (6.4 mm thick) (ft-lb/in. 1/4 in.)	Ease of processing	Approximate relative volumetric cost
ABS	1.01–1.08	91–110 (196–230)	2.07 (300)	102 (215)	107–481 (2–9)	2	2.7
POM	1.41–1.42	160–181 (320–358)	2.76–3.58 (400–520)	110–124 (230–255)	53–107 (1–2)	3	5.2
ETFE	1.7	270 (518)	0.83 (120)	71 (160)	(No break)	9	30
PA 6	1.12	210–220 (410–428)	2.62 (380)	68 (155)	53 (1)	3	4.5
6/6	1.14	255–265 (491–509)	2.89 (420)	75 (167)	53 (1)	3	5.0
6/10, 6/12	1.08	275–310 (383–428)	2.00 (290)	82 (180)	80 (1.5)	3	7.3
PPA	1.17	275–310 (527–590)	1.89–3.27 (275–475)	77–131 (170–268)	43–187 (0.8–3.5)	5	7.9–11.5
LCP	1.4	280–421 (536–790)	13.1–16.5 (1900–2400)	180–355 (356–671)	134–427 (2.5–8)	8–9	37–53
PC	1.21	140–150 (284–302)	2.38 (345)	132 (270)	801 (15)	5	4.9
Polyimide	1.33	300 (572)	3.00 (435)	238 (460)	91 (1.7)	10	49

PPE/PS Blend	1.04–1.10	100–142 (212–288)	2.45 (355)	107 (225)	267 (5)	7	3.7
PEK/PEEK	1.30–1.32	323–381 (613–718)	3.10–3.86 (450–560)	160–166 (320–330)	85–150 (1.6–2.8)	10	66–92
Aliphatic PEK	1.24	220 (428)	1.70 (247)	105 (221)	214 (4.0)	4	16
PPS	1.35	285–290 (545–554)	3.31 (480)	135 (275)	<27 (<0.5)	6	8.0
PP	0.90–0.91	160–175 (320–347)	1.55 (225)	54 (130)	27 (0.5)	2	1.0
PS	1.04–1.05	165–230 (165–230)	3.27 (475)	94 (202)	21 (0.4)	1	1.2
PTFE	2.16	366 (690)	0.34–0.62 (50–90)	46 (115)	107 (2)	10	42
PSU	1.24	190 (374)	2.38 9346)	150 (302)	64 (1.2)	8	51
PVC	1.30–1.45	75–105 (167–221)	2.41–4.13 (350–600)	60–77 (140–170)	21–1175 (0.4–22)	7	2.6
SAN	1.06–1.08	120 (248)	3.86 (560)	104 (220)	27 (0.5)	2	2.1
PBT–PET	1.29–1.40	220–267 (428–513)	1.93–2.76 (280–400)	21–85 (70–185)	27–53 (1–0.5)	3–4	4.3
TPU	1.2	120–160 (248–320)	0.62 (90)	88 (190)	801 (15)	3	4.7

*Estimated relative ease of molding moderately complex part to close tolerances; 10 = hardest, 1 = easiest.

Table 2.4 Representative Properties of Major Thermoplastic 30% Glass Fiber Reinforced Composites

Type	Specific gravity	Tensile strength MPa (Kpsi)	Tensile modulus GPa (Kpsi)	Deflection temperature °C @ 1.82 MPa (°F @ 264 psi)	Notched Izod impact strength J/m (5.4 mm thick) (ft-lb/in., 1/4 in.)	Approximate relative volumetric cost
ABS	1.29	100 (14.5)	7.58 (1100)	102 (215)	69 (1.3)	1.4
POM (c.c.)	1.63	134 (19.5)	9.65 (1400)	166 (330)	96 (1.8)	2.2
ETFE	1.89	97 (14.0)	7.24 (1050)	238 (460)	320 (6.0)	19
PA 6	1.37	159 (23.0)	8.27 (1200)	216 (120)	123 (2.3)	1.9
6/6	1.38	179 (26.0)	8.96 (1300)	254 (490)	107 (2.0)	2.7
6/6 (f.r.)	1.59	134 (19.5)	758 (1100)	243 (470)	69 (1.3)	3.3
6/10, 6/12	1.30	144–152 (21.0–22.0)	7.58 (1100)	213–216 (115–120)	128 (2.1)	1.0
PPA	1.43	221 (32.0)	11.4 (1650)	285 (545)	118 (2.2)	6.2
LCP	1.47–1.67	117–207 (16.9–30.0)	12.8 (1850)	318 (605)	85 (1.6)	17–46
PC	1.41	128 (18.5)	9.45 (1375)	149 (300)	187 (3.5)	3.3
Polyimide	1.43	168 (33.8)	9.70 (2815)	248 (478)	1121 (2.1)	23
PPE/PS Blend	1.27–1.36	128 (18.5)	8.27 (1200	154 (310)	91–107 (1.7–2.0)	2.8

PEK/PEEK	1.47–1.54	176 (25.5)	9.65 (1400)	299 (570)	128.2 (2.1)	30–12
Aliphatic PEK	1.46	130 (18.9)	7.58 (1100)	218 (424)	139 (2.6)	20
PPS	1.56	138 (20.0)	11.0 (1600)	260 (500)	75 (1.4)	4.5
PP (c.c.)	1.14	93 (13.5)	5.52 (800)	149 (300)	102 (0.9)	1.0
PS	1.29	93 (13.5)	8.96 (1300)	102 (215)	53 (1.0)	1.1
PSU	1.48	152 (22.0)	8.96 (1300)	185 (365)	80 (1.5)	3.4
PTFE	2.25	13–16 (1.9–2.3)	1.31 (190)	NA	171 (3.2)	37–41
PVC (20% g.f.)	1.47	128 (18.5)	5.69 (825)	77 (170)	75 (1.4)	2.1
SAN	1.31	120 (17.4)	11.0 (1600)	162 (215)	53 (1.0)	1.4
PBT	1.50	131 (19.0)	10.0 (1450)	218 (125)	80 (1.5)	1.9
PBT (f.r.)	1.63	138 (20.0)	10.3 (1500)	204 (400)	75 (1.4)	2.4
PBT (i.m.)	1.51	124 (18.0)	6.81 (1000)	213 (415)	160 (3.0)	2.0
PET	1.62	152 (22.0)	9.65 (1400)	224 (135)	107 (2.0)	1.8
TPU	1.46	57 (8.2)	1.31 (190)	77 (170)	507 (9.5)	4.2

Note: "g.f." = glass fiber: "f.r." = also flame retardant additives; "c.c." = chemically coupled; "i.m." = impact modified.

Table 2.5 Comparative Properties of Plastics Reinforced with Glass vs. Other Fibers

Base polymer	Fiber, % by weight		Tensile strength MPa (Kpsi)	Compressive strength MPa (Kpsi)	Tensile modulus GPa (Kpsi)	Deflection temperature °C @ 1.82 MPa (°F @ 264 psi)	Notched Izod impact strength J/m (6.4 mm thick) (ft-lb/in. (1/4 in.))	Specific gravity	Approximate relative volumetric cost
PA 6/6	Glass	30	186 (27)	186 (27)	8.96 (1,300)	254 (190)	112 (2.1)	1.37	1.0
	"Long" glass"*	30	193 (28)	193 (28)	10.00 (1,450)	255 (491)	182 (3.4)	1.37	1.0
	Carbon	30	241 (35)	200 (29)	20.00 (2,900)	256 (492)	91 (1.7)	1.28	4.2
	Glass	20	131 (19)	159 (23)	5.86 (850)	254 (490)	64 (1.2)	1.20	0.9
	Carbon	20	193 (28)	172 (25)	16.50 (2,400)	256 (492)	59 (1.1)	1.23	3.1
	Carbon (pitch)	20	90 (13)	138 (20)	10.30 (1,500)	232 (450)	32 (0.6)	1.24	3.0
	Aramid	20	103 (15)	—	4.83 (700)	249 (480)	53 (1.0)	1.28	4.3
	Nickel coated carbon	20	103 (15)	—	9.65 (1,400)	241 (465)	43 (0.8)	1.25	5.0
PC	Glass	20	110 (16)	110 (16)	5.86 (850)	149 (300)	182 (3.4)	1.34	1.0

Carbon	20	131 (19)	131 (19)	13.79 (2,000)	149 (300)	107 (2.0)	1.28	2.5
Aramid	20	83 (12)	—	4.07 (590)	138 (280)	43 (0.8)	1.25	5.0
Stainless steel	10	76 (11)	76 (11)	3.79 (550)	141 (285)	85 (1.6)	1.35	2.0
PP								
Glass	40	76 (11)	69 (10)	6.89 (1,000)	149 (300)	96 (1.8)	1.22	1.0
Glass (chem. coupled)	40	103 (15)	69 (10)	6.89 (1,000)	154 (310)	107 (2.0)	1.22	1.1
Asbestos (anthophylite)	40	35 (5)	41 (6)	4.48 (650)	96 (205)	27 (0.5)	1.24	NA
Phenolic								
Milled glass	30	55 (8)	117 (17)	14.48 (2,100)	246 (475)	53 (1.0)	1.72	1.0
Glass	40	97 (14)	228 (33)	18.96 (2,750)	250 (482)	91 (1.7)	1.72	1.0
Carbon	50	124 (18)	290 (42)	44.13 (6,400)	249 (480)	107 (2.0)	1.50	4.0
Aramid	40	159 (23)	152 (22)	20.68 (3,000)	249 (480)	182 (3.4)	1.30	4.0
α-Cellulose	50	45 (6.5)	214 (31)	8.96 (1,300)	177 (350)	59 (1.1)	1.42	0.5

Note: Glass fiber is E-glass, carbon fiber is PAN, unless noted otherwise. Fiber lengths before processing are ≤ 6 mm (1/4 in.) length (* is 13 mm (1/2 in.).

are limited by their low mechanical properties, high cost, and the difficulty of processing. The most commercially important resin is poly-tetrafluoroethylene (PTFE) which, because of its extremely high molecular weight, will not flow when heated. PTFE must be processed by cold forming parts and then heating through the gel point in an oven, in similar manner to powdered metal sintering. It can also be processed by ram extrusion, a technique that does not require the material to flow. Scrap PTFE can be reprocessed, but each thermal history reduces the polymer's molecular weight substantially and, consequently, its mechanical strength properties, by 30% or more. Glass fiber reinforced reprocessed PTFE shows a similar diminishment of properties versus virgin PTFE compounds.

Fluoropolymers that are melt processable, include:

- fluorinated ethylene–propylene copolymer (FEP)
- perfluoroalkoxy (PFA)
- ethylene–tetrafluoroethylene copolymer (ETFE)

These materials trade off varying amounts of temperature resistance for processability and cost. Unlike PTFE, melt processable fluoropolymers can be reprocessed with minimal property loss and compounds based on such recycled materials can be cost effective in certain applications, for example, where the primary requirement is chemical resistance. Other members of the melt processable fluoropolymer family, such as polyvinylidene fluoride (PVDF) and polychlorotrifluoroethylene (PCTFE), are very rarely used commercially in reinforced composites.

2.2.1.4   Liquid Crystal Polymers

Liquid crystal polymers (LCP) are a relatively unique class of partially crystalline aromatic polyesters based on p-hydroxybenzoic acid and related monomers. They exhibit a highly ordered alignment of the polymeric chains during flow in the "melt" state, which is retained in the cooled, solid state. LCPs possess outstanding mechanical properties over a wide range of temperatures, high heat and chemical resistance, excellent dimensional stability, and inherent flame retardancy. Depending on type, they range in processability from moderately easy to difficult (because of their high molding temperature requirements, although their rheology is such that they tend to flow readily). Unfortunately, ease of processability tends to vary inversely with temperature resistance. Their primary limitation is reduced mechanical properties at right angles to the direction of polymer chain alignment. Their cost is comparable to fluoropolymers on a volumetric basis, varying roughly in proportion to the level of temperature resistance.

Development work is now in progress to utilize LCPs as fibrous reinforcements, primarily in polyolefin or thermoplastic polyester matrices. The novelty of this approach lies in the idea that the composite may be recycled many more times than is the case with conventional fiber reinforcement, because the LCP fibers can regenerate in length each time they are remelted, rather than undergo the usual attrition.

2.2.1.5 Nylons

Nylons or polyamides (PAs) fall into two major subtypes—nylon 6 and 6/6—plus a number of specialty subtypes such as 4/6, 6/10, 6/12, 12/12, 11, and 12. Other specialty subtypes include 6–6/6 copolymers, which have impact and temperature resistance intermediate between their homopolymer parents, and "amorphous" and "aromatic" grades, which contain various aromatic monomers for the purpose of increasing temperature resistance and reducing water absorption. The nomenclature of these normally crystalline materials is taken from the monomer(s) from which they are made; for example, nylon 6 comes from caprolactam and nylon 6/6 from hexamethylenediamine (HMD) and adipic acid. HMD and sebacic acid (made from castor oil) are used in 6/10; HMD and dodecanoic acid are used in 6/12. Nylons are tough and wear resistant, have good chemical resistance, but will absorb moisture, particularly 6 and 6/6.

The first group of specialty types, for example, 6/10, 6/12, 12/12, 11, 12, are less water absorbent than 6 or 6/6 by virtue of their long hydrophobic paraffinic chains between the hydrophilic amide groups. Nylon 4/6 is made from 1,4-diaminobutane and adipic acid and exhibits many properties comparable to nylon 6/6. Although it can absorb more moisture than nylon 6/6, it has generally better heat and creep resistance, toughness, and fatigue endurance; it is also significantly more expensive.

Nylons exhibit low coefficient of friction, good dielectric properties, and excellent fatigue resistance. Their excellent processability and adhesion to reinforcements and fillers make them natural candidates for high loadings of modifiers. Good formulated flame retardant grades are available.

2.2.1.6 Polycarbonate

Polycarbonate (PC), an amorphous carbonic acid polyester, loses two of its outstanding properties when reinforced—transparency and toughness—but its good dimensional stability and superior creep (deformation under continuous load) resistance are much enhanced. It has excellent dielectric and flame retardant properties but only fair chemical resistance. Polycarbonate may be blended with other polymers, such as PBT and ABS. Overall, it offers a good balance of properties at reasonable cost. Polycarbonate copolymers are presently under development that offer improved temperature resistance.

2.2.1.7 Polyetherketones

Polyetherketone (PEK) and polyetheretherketone (PEEK) are semicrystalline thermoplastic aromatic polyesters that display excellent mechanical and dielectric properties even at sustained high temperatures (upwards of 150 °C). They are inherently flame retardant and generate very low smoke emission if burned. PEK/PEEK show outstanding resistance to hydrolysis and chemical attack. They are relatively costly and require special high temperature molding and extrusion techniques. A new type of PEK, based on carbon monoxide chemistry, has recently been announced; it offers most of these same attractive high properties but at a somewhat lower cost.

2.2.1.8 Polyethylene

Polyethylene comes in a range of densities, 0.917 to 0.965, but "high"-density (HDPE) is the primary resin matrix for fiber reinforced compounds. HDPE is a highly linear, crystalline polymer, whereas low- and medium-density polyethylenes (LDPE, MDPE) are more branched and therefore less crystalline. Polyethylene is also produced in both homopolymer and copolymer form, with α-olefin comonomers used to modify the homopolymer properties, usually to enhance environmental stress-crack resistance. HDPE offers a good balance of chemical resistance, modest mechanical properties, and low cost. Polypropylene is generally preferred over HDPE in most applications because of it has almost all of HDPE's attractive properties plus significantly better temperature resistance.

2.2.1.9 Thermoplastic Polyimides

Thermoplastic polyimide copolymers can be injection molded but have heat resistance approaching that of the thermoset polymer. Polyamide–imide (PAI) and polyetherimide (PEI) exhibit excellent dimensional and thermal stability. Their chemical resistance is also outstanding, although PAI will absorb some moisture. Both materials are costly but inherently flame retardant.

2.2.1.10 Polyphenylene Ether

Polyphenylene ether (PPE), also called polyphenylene oxide, is usually blended with polystyrene or high-impact polystyrene, with which it is compatible, to improve its toughness and processability. The blend offers an excellent balance of properties—rigidity, dimensional stability, heat resistance—when reinforced. This amorphous material is often used in the reinforced state to improve its tendency to stress crack. Although its water absorption is extremely low, it is attacked by many halogenated and aromatic solvents. Other PPE alloys are available, such as blends containing PPS or nylon 6/6, offering improved toughness and chemical resistance. Phenylene-based materials are readily rendered flame retardant by additives.

2.2.1.11 Polyphenylene Sulfide

Polyphenylene sulfide (PPS), a semicrystalline polymer, offers excellent dimensional stability, outstanding heat resistance (over long periods of time), and chemical inertness. It is inherently flame retardant and, if ignited, shows low smoke generation and flame spread values. Reinforcement enhances its otherwise low impact strength. PPS is seldom used for moldings in the unreinforced state. PPS requires high temperature processing.

2.2.1.12 Polypropylene

Polypropylene (PP) is available as a crystalline homopolymer with an excellent balance of chemical and heat resistance, low density, and low unit cost. Copolymers (with a low percentage of ethylene added during polymerization) are also produced to improve toughness, but at some loss in rigidity and heat resistance; nevertheless, they are more popular for reinforced grades. Mechanical properties of PP are moderate. Flame retardant grades are also available. Although not generally classed as an "engineering plastic," reinforced PP often competes successfully with more expensive base resin systems in many low-stress, ambient temperature applications.

2.2.1.13 Polystyrene

Polystyrene (PS), an amorphous, glassy polymer, is rigid, easy to process, and low in cost. It is attacked by most solvents, has low heat resistance, and is brittle. Polybutadiene rubber modified grades are used to upgrade toughness of the base polymer, but are not generally cost effective as the resin matrix in composites. Additives are necessary to render PS grades flame retardant, and the polymer does exhibit good dielectric properties. Where the only significant design requirements are dimensional stability and stiffness at indoor temperatures, PS compounds are worth considering.

2.2.1.14 Polysulfone and Polyethersulfone

Polysulfone (PSU) and polyethersulfone (PES) are high-cost, temperature-resistant, rigid, amorphous materials with low moisture absorption. Reinforcement improves toughness and further enhances dimensional stability, but turns these transparent materials opaque. Their highly aromatic chemical structure makes sulfone polymers inherently flame retardant with low smoke generation.

2.2.1.15 Polyvinyl Chloride

Polyvinyl chloride (PVC) balances low cost, excellent dielectric and chemical resistance, against modest mechanical properties and, in particular, low heat resistance. PVC composites have a narrow processing temperature range and require careful processing to avoid thermal degradation. A normally amorphous polymer, PVC's chlorine content renders it inherently flame retardant.

2.2.1.16 Styrene–Acrylonitrile

Styrene–acrylonitrile (SAN), another amorphous polymer, is rigid, dimensionally stable, and easily processed. It loses transparency when reinforced but gains toughness. Its cost, moisture, and chemical resistance are intermediate between PS and ABS. Flame retardant grades are available.

2.2.1.17 Styrene Maleic Anhydride

Styrene maleic anhydride (SMA), an amorphous polymer, is about 30% less rigid but has two to three times higher notched Izod values than SAN. The maleic anhydride comonomer bonds very well with glass fiber. SMA costs about 20% more than SAN. Flame retardant grades are not offered commercially, but more from lack of market interest than technical difficulty.

2.2.1.18 Thermoplastic Polyesters

Thermoplastic polyesters, mainly polybutylene terephthalate (PBT) and polyethylene terephthalate (PET), show excellent heat resistance (with PET the slightly higher of the two) and dielectric properties, low moisture absorption, and good chemical resistance. Reinforcement improves dimensional stability and toughness. Neither are suitable for hot water service. PET requires higher molding temperatures than PBT. Thin-wall parts must be carefully designed to avoid warpage often associated with post-mold shrinkage in PBT and PET; PET is the most difficult to deal with in this respect because it is amorphous if cooled rapidly after molding and, if annealed at 120 to 221 °C (248 to 428 °F), will slowly crystallize with up to a 9% change in density, and attendant effects on part dimensions. PBT is crystalline at molding. Flame retardant grades are available. A specialty polyester, polycyclohexanedimethanol terephtalate (PCT), has higher temperature resistance than PET or PBT but costs more and has a narrower processing temperature range.

2.2.1.19 Thermoplastic Polyurethanes

Thermoplastic polyurethanes (TPUs), both ether- and ester-based, are elastomeric in nature, extremely tough and abrasion resistant. Esters are usually tougher and more temperature resistant than ethers, but are more prone to hydrolyze and degrade on prolonged exposure to water. Reinforcement improves rigidity and dimensional stability; flame retardant grades are not yet commercial products. Chemical resistance, especially to hydrocarbons, is excellent, although both types will absorb moisture. Some manufacturers note that, unlike other thermoplastics, TPU grades have a 6- to 12-month shelf life (owing to gradual crosslinking) and should not be kept in stock indefinitely.

2.2.2 Thermosets

Thermosets are polymers whose backbone chains have become crosslinked via a reactive or polyfunctional monomer under heat and pressure during fabrication, into essentially one giant molecule. This critical difference from thermoplastics leads to generally better dimensional stability, and heat and chemical resistance on the part of thermosets. Conversely, it leads generally to slower processing times and often unreusable scrap. Thermosets are often brittle without filling or reinforcement and are, therefore, normally charac-terized as compounds rather than as pure resins; epoxies and silicones are exceptions. Table 2.6 summarizes the principal thermoset glass reinforced resins and their properties, showing a range of values from manufacturers' data sheets.

The chemistry of thermoset polymerization and crosslinking divides into two basic types: addition and condensation. Addition polymerization gener-ally takes place by the rearrangement of chemical bonds, without the release of byproducts. Condensation polymerization, on the other hand, takes place by the splitting out of byproducts, usually water molecules, from the reac-tion of the monomers, such as in the esterification of an alcohol by an acid. Because the reaction usually takes place at elevated temperatures, water is volatilized; because the molding process involves this reaction taking place inside the mold, fabrication techniques must provide for adequate venting and removal of the water vapors formed.

Not all thermoset resins exhibit measurable glass transition temperatures similar to thermoplastics. Those that do include diallyl phthalates, epoxies, alkyds/polyesters, and urea–formaldehydes. Those that do not include melamines, phenolics, and silicones. Highly crosslinked thermosets are invar-iably amorphous and actually do have a T_g, but it is masked as the degree of crosslinking increases; the ability to detect the T_g is dependent on the sen-sitivity of the instrumentation. When the temperature of the environment

Table 2.6 Representative Properties of Major Thermoset Resins (40% Glass Reinforced)

Type	Specific gravity	Tensile modulus GPa (Kpsi)	Deflection temperature °C @ 1.82 MPa (°F @ 264 psi)	Notched Izod impact strength J/m (6.4 mm thick) (ft-lb/in. (1/4 in.))	Dielectric strength, ST KV/cm (volts/mil)	Approximate relative volumetric cost
Alkyd	2.0	15.9 (2.3)	232 (450)	27–134 (0.5–2.5)	148 (375)	2.0
Allyl (DAP)	1.9	11.0 (1.6)	232 (450)	27–134 (0.5–2.5)	158 (400)	6.0–13.5
Epoxy	1.8	20.7 (3.0)	260 (500)	27–53 (0.5–1.0)	138–158 (350–400)	3.6–7.9
Amino (melamine)	1.7	16.6 (2.4)	177 (350)	32–53 (0.6–1.0)	118–217 (300–550)	1.7
Phenolic	1.8	12.4–20.7 (1.8–3.0)	177–260 (350–500)	27–53 (0.5–1.0)	118–167 (300–425)	1.0–2.0
Polyester	1.9	13.8–24.1 (2.0–3.5)	249 (480)	53–160 (1.0–3.0)	148–167 (375–425)	1.3–3.7
Polyimide	1.6	19.3 (2.8)	316 (600)	27–160 (0.5–3.0)	177 (450)	7.0
Silicone	1.9	13.8 (2.0)	260–482 (500–900)	267–427 (5.0–8.0)	110 (280)	15.0

rises, those resins with measurable glass transition points (less crosslinking) tend to show a greater loss in mechanical properties than those without (more crosslinking).

2.2.2.1 Aminos

Aminos include primarily melamines, but also ureas, which are reacted with formaldehyde to form hard-surfaced condensation polymers resistant to heat and most chemicals other than strong acids and bases. They have a wide range of color possibilities. Ureas are inherently flame retardant and melamines can be formulated to be so. Aminos are relatively inexpensive and have good dielectric characteristics.

2.2.2.2 Alkyds and Polyesters

Alkyds and (unsaturated) polyesters polymerize by the addition of dihydric alcohols and dibasic organic acids, such as glycol and phthalic anhydride. Hardening agents such as styrene and divinyl benzene, together with free radical initiators, must be added for crosslinking. They process faster than many other thermosets and offer excellent electrical properties (particularly arc resistance) and heat resistance at modest cost. Alkyds and polyesters can be formulated to be flame retardant and are also available in virtually an unlimited range of colors.

2.2.2.3 Allyls

Allyls (most importantly, diallyl phthalate, DAP, and diallyl isophthalate, DAIP) cure by the addition of allyl alcohol and phthalic (or isophthalic) anhydride, forming a unique polyester capable of crosslinking with additional monomer. They feature outstanding dimensional stability and electrical properties; excellent moldability; as well as excellent chemical, thermal, and moisture resistance. Higher priced than alkyds, allyls also can be formulated to be flame retardant and are available in a wide variety of colors.

2.2.2.4 Epoxies

Epoxies are polymers based on epichlorohydrin addition-reacted with bisphenol-A or dibasic acids, and capable of crosslinking with amines, anhydrides, novolac (phenolic), or other monomers. This family of materials offers high mechanical and electrical properties, good colorability, and excellent heat and temperature resistance. They are moderate in cost and inherently flame retardant because of the high content of crosslinked aromatic rings in the polymer.

Brominated epoxy bisphenol-A resins are available for applications that require still further enhanced flame retardant performance.

2.2.2.5 Phenolics

Phenolics (Novolac) are phenol–formaldehyde condensation polymers with a good balance of heat resistance and mechanical and dielectric properties. They are the lowest cost thermoset resins, but are available only in a very limited range of dark colors. Phenolics are another condensation thermoset and can be formulated to be flame retardant. Phenolics are the oldest and most widely used of all the thermosets.

2.2.2.6 Polyimides

Polyimides are addition polymers of aromatic imide groups. The closely linked aromatic structure provides exceptional heat resistance, excellent flame retardancy, dielectric properties, and toughness. Polyimides have broad chemical resistance, being attacked only by bases and concentrated inorganic acids. They are relatively expensive and difficult to fabricate because of the high temperatures required.

2.2.2.7 Silicones

Silicones are a unique class of polymers with the usual carbon-to-carbon polymer backbone linkage replaced by alternating silicon–oxygen linkages. This is much the same as in silica-based materials, such as mica or quartz, but with the addition of pendant carbon-based organic groups, such as phenyls, alkyls, or fluoroalkyls. Although costly, silicones are not difficult to fabricate. They are useful where a combination of very high resistance to heat, excellent dielectric properties, and outstanding toughness is required.

References

1. Clegg, W., Collyer, A.A. *Mechanical Properties of Reinforced Plastics,* (1986) Elsevier Applied Science, Barking, Essex
2. Folkes, M.J. *Short Fiber Reinforced Thermoplastics* (1982) Research Studios Press, Letchworth, Herts.
3. Maine, W., et al, "Some Theoretical Aspects of Fibre Reinforced Thermoplastics," presented at the Society of Plastics Engineers Regional Technical Conference, Toronto, Ont., March 1970

3   Materials Selection

Perhaps the most perplexing question a design engineer faces is "How do I select a material from the dozens of resins available?" As with most decisions involving an overwhelming number of choices, the secret is to narrow the field down to just a few likely alternatives and then study these more closely before making a final selection. In this chapter, the author describes a chart system that provides a simple, quick approach for choosing among different glass reinforced thermoplastic and thermoset composites and that shows how to compare overall differing materials with each other. Because it is difficult to reduce these two material classes to a common table, we recommend choosing several products from each category and then comparing them more closely. This technique is not exhaustive in any sense; it is merely a simplified tool to help screen through dozens of materials to arrive at a few potential candidates. Selection via computerized databases is clearly another viable method, and this process is also described later in this chapter and the following one. However, one does lose some of the perspective on the overall balance, as well as the difference, of properties offered by seemingly similar materials when limited to looking at the data through the computer program's interface. For this reason, the author prefers to do the initial screening on paper which affords one more of a visual opportunity to grasp the comparative differences and similarities between materials. Serendipity is more likely to emerge this way.

The author has assumed that the reader has some idea of which properties the application will need. If this is not the case and the reader finds it initially difficult to select which material properties are important, then it may be more useful to read Chapter 4 on design and then return to this chapter. Chapter 4 addresses the relationship of properties to design requirements in some detail. One must also consider whether the application will require that the material to be used meets a specific automotive or government specification, whether an Underwriters Laboratory listing is required, or possibly FDA or USP compliance. These considerations will help at least to narrow the range of potential suppliers of materials.

3.1   Thermoplastics

Table 3.1 groups polymer families on the *y*-axis and design criteria on the *x*-axis, assigning comparative ratings to each category, from 1 (most desir-

Table 3.1   Glass Reinforced Thermoplastic Compound Selector

Each cell shows the group rating (large number) followed by the individual-material rankings (small numbers, in material order).

G/R Resin Groups	Strength & Stiffness	Toughness	Short Term Heat Resistance	Long Term Heat Resistance	Environmental Resistance	Dimensional Accuracy In Molding	Dimensional Stability	Wear & Frictional Properties	Cost
Styrenics (ABS, SAN, Polystyrene)	3 (2,1,3)	6 (1,2,3)	6 (1,2,3)	6 (1,2,3)	6 (1,2,3)	1 (3,1,2)	5 (2,1,3)	6 (3,1,2)	2 (3,2,1)
Olefins (Polyethylene, Polypropylene)	5 (2,1)	4 (2,1)	4 (2,1)	5 (2,1)	3 (2,1)	5 (1,1)	5 (1,1)	3 (2,1)	1 (1,2)
Other Crystalline Resins — Nylons (6, 6/6, 6/10 6/12, Polyester, Polyacetal)	1 (2,1,3,4,5)	1 (2,3,1,4,5)	2 (2,1,3,2,5)	4 (2,1,3,1,2)	4 (5,4,3,2,1)	4 (1,2,2,2,3)	4 (4,3,2,1,2)	2 (3,2,3,4,1)	3 (1,2,4,1,1)
Arylates (Modified PPO, Polycarbonate, Polysulfone, Polyethersulfone)	3 (4,2,2,1)	2 (3,1,2,3)	3 (4,3,2,1)	3 (4,3,2,1)	5 (3,4,2,1)	1 (4,1,2,3)	2 (4,3,2,1)	4 (4,3,1,2)	4 (1,2,3,4)
High Temp. Resins (PPS, Polyamide-imide)	2 (1,2)	4 (2,1)	1 (2,1)	1 (2,1)	2 (1,2)	4 (1,2)	1 (2,1)	4 (2,1)	5 (1,2)
Fluorocarbons (FEP, ETFE)	6 (2,1)	2 (1,2)	2 (2,1)	1 (1,2)	1 (1,2)	6 (2,1)	6 (2,1)	1 (1,2)	6 (2,1)

Ratings: 1—most desirable; 6—least desirable; large numbers indicate group classification, small numbers are for the specific resins within that group. **Strength and stiffness:** The ability to resist instantaneous applications of load while exhibiting a low level of strain. Materials that demonstrate a proportionality between stress and strain have been assigned better relative ratings. **Toughness:** The ability to withstand impacting at high strain rates. **Short-term heat resistance:** The ability to withstand exposure to elevated temperatures for a limited period of time without distortion. **Long-term heat resistance:** The ability to retain a high level of room-temperature mechanical properties after exposure to elevated temperature for a sustained period. **Environmental resistance:** The ability to withstand exposure to solvents and chemicals. **Dimensional accuracy in molding:** The ability to produce warp-free, high-tolerance moded parts. **Dimensional stability:** The ability to maintain the molded dimensions after exposure to a broad range of temperatures and environments. **Wear and frictional properties:** The ability of the plastic to resist removal of material when run against a mating metal surface. The lower the frictional values, the better the relative rating. **Cost:** The relative cost per cubic inch.

Source: Courtesy of LNP Engineering Plastics, Inc.

able) to 6 (least desirable). The ratings are assessed at a common level of 30% glass fiber reinforcement.

Starting with Table 3.1, first draw up a worksheet with the principal polymer names and design criteria. If you are uncomfortable leaving any out, then assign relative weights to establish importance. For example, if "toughness" is five times more important than all other criteria, and the rest are of about equal importance, multiply the "toughness" selection by five and use the unweighted values for the remainder. Nevertheless, it is better to keep criteria as few and uncomplicated as possible. Some minor adjustments for lesser criteria will be possible later by a choice of reinforcement loading.

Second, transfer to the worksheet the boldface numerical rating in each criteria column you have selected. For example, if "toughness" is a criterion, you would write down "6, 4, 1, 2, 4, 2" from top to bottom on your worksheet in the "toughness" column.

Third, add up the numbers across in each row to the "point subtotal" column. The resin family with the lowest point subtotal is the best for your application on a performance basis.

Fourth, add in the cost factor—weight this as you please—and obtain the total. The resin group with the lowest number is the best for your application on a cost–performance basis.

Fifth, repeat the previous four steps, but this time use the small numbers in the chart and only in the resin family you have found to be the best. Once again, the specific resin with the lowest total is the best for your application on a cost–performance basis. If the totals for some materials differ by only one or two points, you may wish to defer making a final selection until you have run some prototype tests, as described in Chapter 5.

Two hypothetical examples are shown in Tables 3.2 and 3.3

3.2   Thermosets

The selection of thermoset materials for specific product applications may be undertaken in a fashion similar to the procedure outlined previously for thermoplastic materials. The compound selector sheet has been slightly modified in keeping with the characteristics of thermoset materials (Table 3.4).

A rating values chart (Table 3.5) has been developed to assist in developing rating numbers for relative reinforcement loading levels in a given resin matrix. For example, a phenolic resin containing 60% glass fiber would

Table 3.2 Gasoline-Powered Chain Saw Housing
Design Criteria: Strength and Stiffness, Toughness, and Environmental Resistance

Material Characteristics / Design Criteria	Strength & Stiffness	Toughness	Short Term Heat Resistance	Long Term Heat Resistance	Environmental Resistance	Dimensional Accuracy In Molding	Dimensional Stability	Wear & Frictional Properties	Point Sub Total	Cost	Point Total
G/R Resin Groups	X	X	X		X						
Styrenics ABS SAN Polystyrene	3	6	6		6				21	2	23
Olefins Polyethylene Polypropylene	5	4	4		3				16	1	17
Other Crystalline Resins Nylons 6 6/6 6/10, 6/12 Polyester Polyacetal	**1** 2 1 3 4 5	**1** 2 3 1 4 5	**2** 2 1 3 2 5		**4** 5 4 3 2 1				**8** 11 9 10 12 16	**3** 1 2 4 1 1	**11** 12 11 14 13 17
Arylates Modified PPO Polycarbonate Polysulfone Polyethersulfone	3	2	3		5				13	4	17
High Temp. Resins PPS Polyamide-imide	2	4	1		2				9	5	14
Fluorocarbons FEP ETFE	6	2	2		1				11	6	17

Comments: At the point subtotal, "Other Crystalline Resins" and "High Temp. Resins" are contenders. With cost factored in, High Temp. Resins drop from the picture. Rating within the group says, "Use 6/6 or 6 Nylon."
Source: Courtesy of LNP Engineering Plastics, Inc.

Table 3.3   Impeller for Chemical Handling Pump
Design Criteria: Strength or Stiffness, Toughness, Short Term and Long Term Heat Resistance and Environmental Resistance

Material Characteristics / Design Criteria	Strength & Stiffness	Toughness	Short Term Heat Resistance	Long Term Heat Resistance	Environ-mental Resistance	Dimensional Accuracy In Molding	Dimensional Stability	Wear & Frictional Properties	Point Sub Total	Cost	Point Total
G/R Resin Groups	X		X	X	X						
Styrenics ABS SAN Polystyrene	3		6	6	6				21	2	23
Olefins Polyethylene Polypropylene	5		4	5	3				17	1	18
Other Crystalline Resins Nylons 6 6/6 6/10, 6/12 Polyester Polyacetal	1		2	4	4				11	3	14
Arylates Modified PPO Polycarbonate Polysulfone Polyethersulfone	3		3	3	5				14	4	18
High Temp. Resins PPS Polyamide-imide	[1] 2		[2] 1	[2] 1	1 [2]				[6] 6	[1] 5 [2]	[7] 8 11
Fluorocarbons FEP ETFE	6		2	1	1				10	6	16

Comments: "High Temp. Resins" clear choice throughout. Final selection would be PPS based on price advantages. If Heat Resistance had not been a factor, Olefins would have been the choice.
Source: Courtesy of LNP Engineering Plastics, Inc.

Table 3.4   Thermoset Compound Selector Worksheet

Material Characteristics / Design Criteria G/R Resin Groups	Strength & Stiffness	Toughness	Short Term Heat Resistance	Long Term Heat Resistance	Environmental Resistance	Dimensional Accuracy In Molding	Dimensional Stability	Wear & Frictional Properties	Point Sub Total	Cost	Point Total
Aminos	3	3	3	4	3	3	4	3		3	
Alkyds/polyesters	4	1	3	5	2	2	3	5		1	
Allyls (DAP)	5	2	2	2	2	2	2	4		3	
Epoxies	1	3	2	2	1	2	2	3		2	
Phenolics	1	2	1	3	2	1	1	4		1	
Polyimides	3	3	1	1	2	2	2	3		4	
Silicones	4	3	1	1	3	3	3	3		5	

Ratings: 1—most desirable; 6—least desirable; large numbers indicate group classification, small numbers are for the specific resins within that group. **Strength and stiffness:** The ability to resist instantaneous applications of load while exhibiting a low level of strain. Materials that demonstrate a proportionality between stress and strain have been assigned better relative ratings. **Toughness:** The ability to withstand impacting at high strain rates. **Short-term heat resistance:** The ability to withstand exposure to elevated temperatures for a limited period of time without distortion. **Long-term heat resistance:** The ability to retain a high level of room-temperature mechanical properties after exposure to elevated temperature for a sustained period. **Environmental resistance:** The ability to withstand exposure to solvents and chemicals. **Dimensional accuracy in molding:** The ability to produce warp-free, high-tolerance moded parts. **Dimensional stability:** The ability to maintain the molded dimensions after exposure to a broad range of temperatures and environments. **Wear and frictional properties:** The ability of the plastic to resist removal of material when run against a mating metal surface. The lower the frictional values, the better the relative rating. **Cost:** The relative cost per cubic inch.
Source: After LNP Engineering Plastics, Inc.

Table 3.5   Thermoset Rating Values Chart Rating Criteria by Glass Fiber Content

Percent glass fiber content	60–70	40–50	2535	15–20	5–10
Flex Modulus, GPa	17.2	13.8	10.3	6.9	3.4
Mpsi	2.5	2.0	1.5	1.0	0.5
Notched Izod, J/m	53	37	27	16	11
ft-lbs/in.	1.0	0.7	0.5	0.3	0.2
Heat Resistance					
Short term: < 1000 h, °C (°F)	260 (500)	204 (400)	149 (300)	93 (200)	NA
Long term: ≥ 1000h, °C (°F)	204 (400)	149 (300)	121 (250)	NA	NA
Susceptibility to chemical attack	Inert	Slight	Moderate	Severe	Severe
Dimensional tolerance in mldg,					
cm/cm	±0.003	±0.007	±0.013	±0.018	±0.018
in./in.	±0.001	±0.003	±0.005	±0.017	±0.007
Wear susceptibility	Slight	Moderate	Severe	Severe	Severe
Relative volumetric cost ratios	1	2	3	6	10

exhibit a flex modulus of 17.2 GPa (2.5 Mpsi) whereas a phenolic compound with 30% glass would yield a flex modulus of 10.3 GPa (1.5 Mpsi).

Thus, the selection criteria on the worksheet may be adjusted to take into account the effect on properties of glass fiber content in any given resin matrix.

The same hypothetical chain saw housing and chemical pump example applications may be used again, this time to show how thermoset materials compare in use; the thermoset selector worksheets are shown in Tables 3.6 and 3.7.

The same procedure as drawn for thermoplastics can be utilized for thermoset selection, that is, the lowest point total represents the most appropriate material to be considered. Adjustment of the point total can be made by selection of modifications (fiber content) in the resin matrix category for further definitions of the precise compound. It is at this point that consultations with molding material suppliers are desirable to obtain assistance with the final selection.

3.3   Using a Computer for Selection

As mentioned previously, one can use the computer as a tool to screen materials for a design. The procedure is similar to the paper exercise. Again, draw up a list of the critical properties for the application, but this time set the

Table 3.6   Chain Saw Housing; Compound Selector Worksheet: Thermoset

Material Characteristics — G/R Resin Groups	Strength & Stiffness	Toughness	Short Term Heat Resistance	Long Term Heat Resistance	Environ-mental Resistance	Dimensional Accuracy In Molding	Dimensional Stability	Wear & Frictional Properties	Point Sub Total	Cost	Point Total
Design Criteria	X	X	X		X						
Aminos	3	3	3		3				12	3	15
Alkyds/polyesters	4	1	3		2				10	1	11
Allyls (DAP)	5	2	2		2				11	3	14
Epoxies	1	2	2		1				6	2	8
Phenolics	1	2	2		2				7	1	8
Polyimides	3	3	1		2				8	4	12
Silicones	4	3	1		3				11	5	16

Ratings: 1—most desirable; 6—least desirable; large numbers indicate group classification, small numbers are for the specific resins within that group. **Strength and stiffness:** The ability to resist instantaneous applications of load while exhibiting a low level of strain. Materials that demonstrate a proportionality between stress and strain have been assigned better relative ratings. **Toughness:** The ability to withstand impacting at high strain rates. **Short-term heat resistance:** The ability to withstand exposure to elevated temperatures for a limited period of time without distortion. **Long-term heat resistance:** The ability to retain a high level of room-temperature mechanical properties after exposure to elevated temperature for a sustained period. **Environmental resistance:** The ability to withstand exposure to solvents and chemicals. **Dimensional accuracy in molding:** The ability to produce warp-free, high-tolerance moded parts. **Dimensional stability:** The ability to maintain the molded dimensions after exposure to a broad range of temperatures and environments. **Wear and frictional properties:** The ability of the plastic to resist removal of material when run against a mating metal surface. The lower the frictional values, the better the relative rating. **Cost:** The relative cost per cubic inch.

Source: After LNP Engineering Plastics, Inc.

Table 3.7   Compound Selector Worksheet: Thermoset

Material Characteristics / G/R Resin Groups	Strength & Stiffness	Toughness	Short Term Heat Resistance	Long Term Heat Resistance	Environmental Resistance	Dimensional Accuracy In Molding	Dimensional Stability	Wear & Frictional Properties	Point Sub Total	Cost	Point Total
Design Criteria	X		X	X	X						
Aminos	3		3	4	3				13	3	16
Alkyds/polyesters	4		3	5	2				14	1	15
Allyls (DAP)	5		2	2	2				11	3	14
Epoxies	1		2	2	1				6	2	8
Phenolics	1		2	3	2				8	1	9
Polyimides	3		1	1	2				7	4	11
Silicones	4		1	1	3				9	5	14

Ratings: 1—most desirable; 6—least desirable; large numbers indicate group classification, small numbers are for the specific resins within that group. **Strength and stiffness:** The ability to resist instantaneous applications of load while exhibiting a low level of strain. Materials that demonstrate a proportionality between stress and strain have been assigned better relative ratings. **Toughness:** The ability to withstand impacting at high strain rates. **Short-term heat resistance:** The ability to withstand exposure to elevated temperatures for a limited period of time without distortion. **Long-term heat resistance:** The ability to retain a high level of room-temperature mechanical properties after exposure to elevated temperature for a sustained period. **Environmental resistance:** The ability to withstand exposure to solvents and chemicals. **Dimensional accuracy in molding:** The ability to produce warp-free, high-tolerance moded parts. **Dimensional stability:** The ability to maintain the molded dimensions after exposure to a broad range of temperatures and environments. **Wear and frictional properties:** The ability of the plastic to resist removal of material when run against a mating metal surface. The lower the frictional values, the better the relative rating. **Cost:** The relative cost per cubic inch.
Source: After LNP Engineering Plastics, Inc.

minimum and maximum values that appear to be needed for the application. Enter these values and then instruct the program to search. Most computer materials database search programs will then sort based on this input and list the materials that meet these criteria. Appendix A.6 lists many of these programs. Comments on just a few of the more popular programs follow.

PLASPEC, an online database, will permit selection of not only of some 50 mechanical, thermal, and electrical properties, but also cost, regulatory listings (e.g., UL, NSF, FDA compliance), chemical resistance, and supplier; the final grouping can be printed out, including all of the property values used in sorting. This type of format provides the broadest range of information, but can be incomplete or dated because it depends on the supplier furnishing current data on a real time basis to the data service company; also, data from different suppliers are not necessarily comparable, because of differing test specimen preparation and testing procedures. Naturally, the service is fee based.

In contrast, RTP Company's proprietary database permits sorting on much the same number and type properties, but the output is in the form of RTP's product grade numbers, which then must be checked against individual data sheets to obtain the actual property data. The GE Select proprietary database will print property data, but does not cover as wide a range of materials. The advantage to these programs is that they are furnished free of charge to the user. The limitation, of course, is that each database includes only the products made by that particular company; again, one set of data is not necessarily comparable to another.

CAMPUS database information is more ideal in that the test data from different suppliers are, in principle, directly comparable. Again, one can define the properties and their values that are of interest, sort, and print. Unfortunately, although the number of suppliers who furnish their data in CAMPUS form is growing, the ones who do are not yet in the majority.

4 Design Considerations

4.1 Basic Concepts

An engineer who is designing a part using a short fiber reinforced plastic naturally begins with the usual considerations. These start with establishing the following:

- Function(s)—What is the part required to do? How may it be shaped?
- Performance—What are the range of conditions over which the part's functions must be performed? What properties of materials correlate with these conditions?
- Economics—What are the dimensions of the part? Its weight? What tooling and machinery will be required to fabricate it and what will be their cost and output? What is the maximum part cost limitation?

Once these elements have been identified, then a preliminary design geometry can be selected, the critical properties determined, studied, compared to the range of properties of available materials, and a preliminary plan for fabrication outlined.

Assuming that the initial elements of form, performance, and economics for a design are in mind, the balance of this chapter will take up specific considerations of the design properties of this class of materials. Although mechanical properties such as strength, stiffness, toughness, and fatigue endurance are often the usual starting points, temperature, time, and environment will strongly influence these values. In addition, some applications will also entail specialized criteria, such as dimensional compatibility (e.g., where parts made of different materials are joined to one another), wear (tribological) situations, or dielectric and flammability considerations. The latter two elements often dictate that the material used be one of those listed by Underwriters Laboratories, thereby narrowing the choices. The design engineer must consider carefully all of these influences on the performance of the function of the part, weigh their relative importance, strike the required balance, and then choose the most economical material on a life cycle basis.

This latter point is a most important one, to ensure that quality considerations are not compromised or overlooked, as well as fabrication and future maintenance costs. Costs should never be oversimplified or allowed to overwhelm the initial design considerations. Short fiber reinforced plastics are relatively expensive on a weight basis, compared to unreinforced plastics or conventional materials. However, their performance often more than compensates for this difference. Some designers may be so put off by the high unit price of one of these materials that they may fail to look beyond it. The price per unit weight of a given material, however, is only one of many points of design economics to be considered and never the sole one, as all too often seems to be the case. Parts are unacceptably expensive no matter what their cost if they do not function as intended, especially if the reason is that they have been made of a less costly but inappropriate material.

In addition, there is a major technical caveat with regard to designing in short fiber reinforced plastics, and that is the effect of orientation (anisotropy): the tendency of the reinforcing fibers to align in the direction of flow, and hence to cause properties to vary directionally. This phenomenon is highlighted again and again because it is so important. It affects strength and modulus particularly and is in turn affected most strongly by part and tool design as well as by processing. Keep in mind as you proceed, orientation can create up to a 40% variation in the properties mentioned (although in typical applications the variation is closer to 10 to 15%). The number of designs that can actually utilize orientation to the advantage of performance are few, for example, fastening bolts that have been molded with longitudinal flow for maximum tensile strength. In most designs, the engineer must be alert for conditions that produce undesirable orientation, and correct them wherever possible.

A great benefit of short fiber reinforced plastics is that they afford opportunities to design components in plastic that otherwise would not be feasible in anything other than metal or similar conventional material. An example might be chemical pump components where a combination of stress and an aggressive chemical environment would preclude the use of unreinforced plastics or all but the most exotic and costly of metals. Nevertheless, the designer should avoid the common trap of merely substituting a plastic material in a design originally prepared for another material, such as steel. This approach does not allow for the differences between the materials, such as ductility and creep resistance, and only guarantees that there will be problems in fabrication and performance.

Owing to the complexity and diversity of potential applications, unfortunately, there is no easy way to distinguish between suitable and unsuitable applications and designs. In this respect, there is no real substitute for experience and judgment. Perhaps more importantly, there is no real substitute for prototyping and testing. Although this book is not meant to be a design manual, it is intended to present an overview of those design-related aspects of properties that are intrinsic and specific to short fiber reinforced plastics, as well as to alert the designer to important features that differ from those of unreinforced plastics and, where unusually significant, conventional materials.

4.2   Computer Assisted Design

Fortunately, today's design engineers have a growing wealth of computerized assistance. Computer databases help with the comparative properties of materials. In particular, an increasing number of polymer producers are now participating in a voluntary program called CAMPUS, standing for *Com*puter *A*ided *M*aterial *P*reselection by *U*niform *S*tandards. A consortium of European producers joined together in 1988 to offer properties data on their materials, based on tightly defined International Standards Organization (ISO) procedures, which provide for uniform test molds and specimen designs, uniform molding and specimen preparation conditions, uniform testing parameters, and a common, self-contained computer operating system interface. The major driving force behind CAMPUS was not just to harmonize software but to provide truly comparable materials property data. American Society for Testing and Measurement (ASTM) protocols, the primary standards used in the United States, allow considerable latitude in these areas that ISO defines so tightly. Some 40 companies at present provide computer diskettes in the CAMPUS format to prospective customers. Although just 13 of these are United States-based producers with English language software, the number is expected to grow rapidly because a number of other United States producers now offer materials data in CAMPUS form in Europe in the German language (and, in a some cases, Spanish or French). Moderately priced software exists to merge and search multiple CAMPUS databases. A number of United States firms have their own proprietary materials databases on diskette, and some also have information available via an Internet home page.

In the area of part design, computer aided design (CAD) allows the designer to envision and revise at will the solid geometry of a part. Computer aided engineering (CAE) programs provide for finite element, mold flow, warpage, and cooling analysis; a growing number are specific to reinforced materials. As the by now wary design engineer will suspect, however, it is well to know that there are boundaries to the accuracy and reliability of CAD/CAE. For example, do not ignore the insulating effect of the first part in a two-shot molding analysis of cooling and filling, or, conversely, the heat sink effect of a metal insert [1].

Another software program, design for manufacture and assembly (DFMA), provides powerful techniques for evaluating part and assembly configurations, together with detailed cost estimates. Computer aided manufacturing (CAM) and computer integrated manufacturing (CIM) programs are the logical extension of CAD to the shop floor.

These computer techniques are still evolving, with the required specialized knowledge and significant investment costs limiting their use at present. Nevertheless, unquestionably no business will be competitive for long without it [2]. As new and more powerful software and hardware come into the market, these tools are becoming rapidly more affordable and accessible. A listing of the more popular current programs is shown in Appendix A.6. In addition, there are services available to help the designer evaluate the utility of different software [3].

As the reader pages through the rest of this chapter, it may be observed that there is a preponderance of design data on thermoplastic materials as compared to thermosets. Although this is undeniably true, it is not because of any subjective bias on the part of the author against thermosets. It is regrettably due to the paucity of thermoset resins design data at present available from their manufacturers.

4.3   Dimensional Stability

One of the advantages of using short fiber reinforced plastics versus unreinforced materials is their reduced dimensional change over a range of temperatures. This is most immediately observed as low mold shrinkage, permitting the production of parts to tighter tolerances than would otherwise be the case. For example, unreinforced nylon 6/6 may have a mold shrinkage of up to 0.2 mm/cm (0.020 in./in.), whereas a 30% glass fiber reinforced nylon 6/6 compound may have a mold shrinkage under the same conditions of only 0.06 mm/cm (0.006 in./in.). Obviously, it will be much easier and less costly to fabricate parts reproducibly to a specified tolerance of plus or minus 1 mil in a material that shrinks only up to 6 mils to begin with, than it would be in a material that shrinks up to 20 mils.

The low shrinkage and elongation of short fiber reinforced plastics carries a caution: one must allow adequate draft angle on part walls to facilitate ejection from the mold. An angle of 2 to 3 degrees generally works well.

Naturally, undercuts that must be stripped from a mold are virtually unworkable in these materials.

The low mold shrinkage of short fiber reinforced plastics also makes possible designs that mate plastic and metal components, such as molded-in metal inserts, without running into problems with significant stresses being created in the combined part during molding.

Also, metal parts can be glued or otherwise fastened to plastic parts, provided the coefficients of thermal expansion match up. For example, the coefficients of thermal expansion per test ASTM D696, of aluminum (die-cast type ASTM S9) and 40% glass fiber reinforced nylon 6 are nearly identical over a range of -30 °C to $+30$ °C (-22 °F to $+86$ °F), at a value of 0.061–0.067 m/m/°C $\times 10^{-6}$ (1.1–1.2 in./in./°F $\times$ KSB10^{K}F^5); Table 4.1 shows this and other metal/plastics property comparisons. It may be noted from Table 4.1 that highly loaded thermoplastic composites can have a lower coefficient of thermal expansion than thermosets. Care must be taken when working with coefficients of expansion to ensure that they correspond to the temperature range under consideration for the application. This is because these values are usually not linear with temperature, particularly above the glass transition temperature of crystalline polymers; the values tend to be significantly higher above the T_g than they are below it.

It is also evident from the strength-to-weight ratios (or specific strength, which is tensile strength divided by specific gravity) shown in Table 4.1 that short fiber reinforced plastics can replace die casting metals at substantial savings in weight and thickness, while providing equivalent, if not superior, performance. Many designers are not aware of the advances in plastics materials that afford actually greater strength than is found in many die casting metals. Hence, where strength is the critical property in choosing wall thickness, a material such as carbon fiber reinforced nylon 6/6 will allow for a thinner, lighter wall than would be required in, for example, magnesium. Where modulus is the critical property, however, the use of short fiber reinforced plastics will normally not allow a thinner wall, but often will allow a lighter one. It is important to note that modulus, rather than strength, is typically the limiting property in most load-bearing applications.

One other note of caution to be observed when determining the dimensional stability of a short fiber reinforced plastic for an application: the sometimes unexpected post-molding dimensional changes. These are usually irreversible and are the result of three causes:

- post-molding relaxation of molded-in stresses,
- post-molding crystallization,
- post-molding curing.

Table 4.1   Short Fiber Reinforced Plastic versus Die Casting Metal Alloys Selected Properties Comparison

Property	ASTM test method	Nylon 6 40% glass fiber	Nylon 6/6 40% carbon fiber	Phenolic 50% glass fiber	Aluminum ASTM S9	Magnesium AZ91B	Zinc ZA-12
Specific gravity	D792	1.46	1.34	1.80	2.89	1.81	6.61
Tensile strength, MPa	D638	179	276	55	117–207	228	283
(Kpsi)		(26)	(40)	(8)	(17–30)	(33)	(41)
Flexural modulus, GPa	D790	10.3	23.4	14.5	68.9	46.9	41.4
(Kpsi)		(1500)	(3400)	(2100)	(10000)	(6800)	(6000)
Impact strength, J/m	D256						
Charpy, unnotched 6.4 mm (1/4 in.) bar (ft-lb/in.)		—	—	—	32–187 (0.6–3.5)	107 (2.0)	2296 (43)
Izod, unnotched 6.4 mm (1/4 in.) bar (ft-lb/in.)		1068 (20)	641 (12)	160 (3)	—	—	—
Coefficient of linear thermal expansion	D696						
(−30 to +30°C) m/m/°C ×10^{-6}		0.067	0.061	0.111	0.067	0.078	0.083
((−22 to +86°F) in./in.| ×10^{-5})		(1.2)	(1.1)	(2.0)	(1.2)	(1.4)	(1.5)
Rockwell hardness	D785	M92	M106	M120	NA	NA	M103
Strength-to-weight ratio*, Mpa		122.6	206.0	30.6	40.5–71.6	126.0	42.8
(Mpsi)		(17.8)	(29.9)	(4.4)	(5.9–10.4)	(18.2)	(6.2)
Modulus-to-weight ratio*, Gpa		7.05	17.46	8.06	23.84	25.91	6.26
(Kpsi)		(1027)	(2537)	(1167)	(3460)	(3757)	(908)

*Strength or modulus divided by specific gravity
Source: LNP Engineering Plastics, Inc.

All can cause uneven part shrinkage and hence warpage; very often the first two problems can often be minimized or eliminated after molding by annealing the part for 15–60 minutes at the temperature recommended by the manufacturer.

Molded-in stresses should be avoided through careful part and tool design to minimize differential cooling. For example, thick part wall sections (particularly those that transition abruptly into thin wall sections), inadequate mold cooling, and undersized gates can all cause differential shrinkage [4]. The solution to the second problem can often be to carefully control molding conditions, and avoid the use of materials that have a tendency toward this problem, such as PET. If the part or mold design is such that some warpage seems unavoidable, some improvement may be possible by using glass fiber and glass bead or mineral filler combinations to reduce anisotropy; also, using high concentrations of fiber, for example, 50, 60, or even 70%, will reduce mold shrinkage values to the point where the difference between flow and transverse directions will be almost nil. Processing conditions do play a major role in controlling shrinkage, and these are treated in more detail in Chapter 7 [5].

The third problem, a thermoset phenomenon, can be avoided by careful selection of materials and processing conditions to ensure complete curing during the molding process.

4.4 Stress, Strain, and Time

Plastics, particularly thermoplastics, exhibit different mechanical behavior under stress than metals, as widely documented in scientific literature. Hooke's Law, which states that strain is directly proportional to stress, cannot be safely applied to plastics. Being viscoelastic rather than truly elastic, plastics tend to continue to deform under long-term loads, even at relatively low temperatures. Where the load is constant, this is called creep; in applications such as gaskets, where the deformation of the material causes the load to diminish, it is called stress relaxation.

Another consequence of viscoelasticity is that the stress–strain curves, and hence derived strength and modulus values, are highly sensitive to the rate of loading. These values are also very sensitive to temperature, but this is taken up in more detail later in this chapter. Although reinforcement reduces creep and stress relaxation, it does not eliminate them. An example of the effect of different lengths of glass fiber reinforcement on tensile creep is shown in Fig. 4.1. Thermosets show much better creep resistance than thermoplastics, because the crosslinked polymer chains cannot slide past each other under load as noncrosslinked thermoplastic ones will.

Although reinforcement increases the static strength and stiffness of the polymer matrix, it also generally reduces dynamic toughness, because elongation is reduced. This in turn means that the nature of fracture failure tends to be of a brittle character rather than ductile. This topic is also covered in more detail later in this chapter.

Reinforcement also improves the resistance of the matrix to cyclic loading failure (fatigue endurance). Carbon fiber reinforced composites have been found to give better fatigue endurance than glass fiber reinforced ones, presumably because of the better thermal conductivity and hence heat dissipation conferred by carbon fiber. Typically, crystalline polymers show distinct fatigue endurance limits: when strength values are plotted against cycles-to-failure, the curve becomes asymptotic. Amorphous polymers generally do not exhibit a fatigue endurance limit, but rather show a declining strength curve vs. cycles-to-failure [6]. Glass fiber reinforced polycarbonate is an exception to this rule, as it exhibits a well-defined fatigue endurance limit. Long glass reinforced compounds show better fatigue resistance than short glass compounds at initial and intermediate cycles, but these differences tend to converge at high cycles, as noted in Fig. 4.2.

In view of the ambiguities noted previously, what is one to do? The design engineer must evaluate the nature of the stresses that will be applied to the part, and whether they are of long duration or repeat frequently. Strength and modulus values must be adjusted downwards to reflect the effects of creep and fatigue, using strength and modulus data that have been developed under conditions more closely aligned with those of the actual application, rather than standard short-term, room temperature test data [7]. Of course, the latter values may well be perfectly suitable if the applied stresses are of short, infrequent application. In general, it is wise to use tensile strength values *at yield* rather than *at break,* because deformation past the yield point is permanent and therefore will usually constitute part failure before rupture actually occurs. For many short fiber reinforced plastics, strength at yield and break are often the same thing, because failure tends to be brittle rather than ductile.

As noted earlier, long glass fiber reinforced thermoplastics show significantly better long term creep resistance than do short fiber reinforced compounds, particularly at elevated temperatures [8]. If the application is subject to significant loads at temperatures above ambient, this

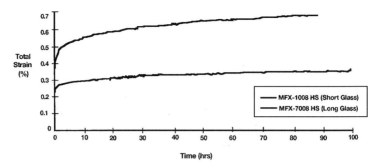

Figure 4.1 Effect of glass length on tensile creep
(Courtesy of LNP Engineering Plastics, Inc.)

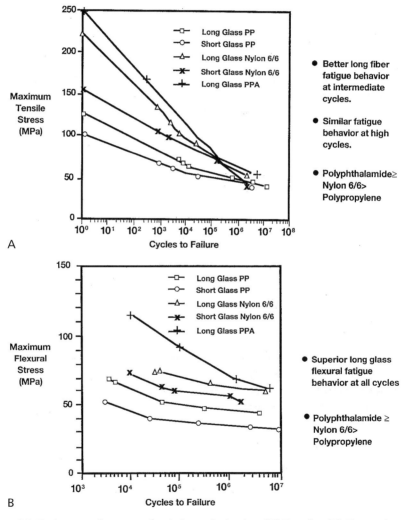

Figure 4.2 Fatigue endurance of reinforced plastics; (A) Tensile; (B) Flexural
(Courtesy of LNP Engineering Plastics, Inc.)

class of materials offers a certain comfort from the standpoint of the more reliable performance they offer. This must be offset, however, against their higher cost. One must weigh whether it is more cost effective to use a more expensive material in a thin wall or a less costly one in a thicker wall.

Another factor that diminishes published strength and toughness values is the presence of pigments in concentrations above one half of one percent. Often overlooked, colorants are really just another filler and can have a disastrous effect on predicted part performance if an intense color has been specified but design calculations were based on unpigmented material. Titanium dioxide is particularly deleterious on glass fiber, as even 0.1% can reduce strength and toughness values by 10 to 20%. Its effect seems to be caused by its abrasiveness on the glass, causing the fibers to break into suboptimum lengths. Fortunately, it is seldom used in high concentrations because of its high hiding power. However, it may be sometimes inadvertently present, for example, as a delustering agent in nylon, if the base resin was originally intended for fiber use.

One more critical factor that must be considered when designing parts, regardless of the material (but often unintentionally overlooked), is the matter of the reliability of published properties. The published data upon which we must rely are averages (statistical means). However, one is reminded of the 6-foot tall statistician who drowned in a 3-foot average depth stream! It is imperative to obtain minimum property values, standard deviation values, or even property value histograms to interpret the data intelligently enough to design parts to meet zero-defect quality standards. Manufacturers' products may exhibit flattened, skewed, or even bimodal distributions of property values, rather than the expected nicely bell-shaped curve. Designing at three or even four standard deviations from the mean require accurate and detailed knowledge of the test data for a given product. Manufacturers' data sheets may also reflect annealing, moisture conditioning, and other conditions that, although within ASTM test boundaries, enhance test values in ways that do not necessarily correspond to other manufacturers data. This is another reason to look for CAMPUS data.

In short glass reinforced materials, a good rule of thumb is "the better the dispersion of the reinforcement, the more reproducible the property value." For this reason, the author strongly recommends against the use of glass fiber concentrates or "dry blends," which are offered by some vendors as a less expensive substitute for fully dispersed, precompounded products. As with many things in life, one usually gets what one pays for, and there are a number of hidden costs in this particular shortcut: sharply increased variability of properties, increased scrap and faulty parts, increased machine and tool wear, and aesthetically poorer part appearance, all caused by glass fibers that are not fully wetted with polymer or are not fully dispersed. These results are the antithesis of quality.

4.5   Temperature

Perhaps the most critical environmental condition that must be considered in plastics design is temperature. Because the physical properties of short fiber reinforced plastics show to greater advantage over unreinforced plastics at elevated temperatures, they are more likely to be considered for uses that involve this parameter. We have discussed the effect of temperature on part dimensions in an earlier section, but this is a minor point compared to the effect of temperature on strength, stiffness, and toughness.

Of course, even the performance of metals is affected by temperature: just as plastics do, metals lose toughness at low temperatures and will creep at high temperatures under load. A rule of thumb is to design in metal for use at no more than two thirds of its melt point; for example, such low-melting metals as aluminum and magnesium die-casting alloys are not considered suitable for use at service temperatures above 315 °C (600 °F).

Low temperatures generally raise strength and stiffness, but usually at the expense of impact resistance. Consequently, applications involving low-temperature impact strength as a key property must often employ a toughened material, generally through the use of an elastomer blended into the composite.

High temperatures raise more complex questions. The first point to be considered is whether the exposure to elevated temperature is for a short or extended period of time. If the time frame is brief (say, for only an hour or less), then heat deflection temperature (HDT) as measured by ASTM D648, although a relatively crude tool, is commonly used to screen candidate materials. However, because this test tells us nothing about the long-term effect of high temperature on a material, it must be used with great caution. HDT at least gives an idea of where a material's short-term load-bearing ability has been reached, which is more important for amorphous than crystalline materials, as mentioned earlier [9]. Such tests as Vicat softening point and melting point provide even less information. When comparing HDTs, the engineer must take care to ensure that the values being used were all obtained under the same loads, because it is common to use 4.55 kPa (66 psi) for lower melting point polymers, such as poly(vinyl chloride) (PVC), styrenics, and polyolefins, and 1.72 MPa (264 psi) for most other polymers.

Another useful test is dynamic mechanical analysis (DMA), which measures stiffness vs. temperature [10]. This information is becoming more available for thermoplastics and is superior to HDT information because it provides a range of data rather than a single point.

Underwriters Laboratories (UL) has adopted its own standard for rating materials used at elevated temperatures for extended periods, the UL temperature index (TI), identified in UL 746B, "Polymeric Materials—Long Term Property Evaluations." UL's TI is a comparative rating of the ability of a material to retain 50% of the initial value of one or more critical properties in the course of extended exposure to elevated temperatures (extrapolated via an Arrhenius plot to a life expectancy of 100,000 hours or more than 11 years), so that applying the typical design safety factor of two, the material will function as intended in an application. The properties involved are primarily, but not necessarily limited to, impact strength (normally the property most affected by long-term exposure to heat), tensile strength, flammability, and dielectric strength. So as not to exclude a material from use under conditions where one of the above properties may not be a factor, more than one TI may be assigned. Note that while the materials are aged at elevated temperatures, the properties are tested at standard room temperature: the test does not represent that the room temperature properties values are the same as at the TI. Although the TI is a useful tool—and, in fact, a virtually mandatory one for electrical applications—its utility is limited to only those materials whose manufacturers have obtained TIs for them.

Parts that will function at high temperatures must be designed using data that have been generated at these or higher temperatures, and not at ambient temperatures. Such data are frequently available from the supplier of the material; examples are shown in Fig. 4.3. Great care should be exercised to ensure that the data used were, in fact, obtained on specimens that had been appropriately aged at the test temperatures; short-term testing offers no assurance whatever that the material will retain these properties over a period of time at the higher temperature. Thermosets tend to perform more predictably than thermoplastics at sustained high temperatures, because they are for the most part less subject to oxidative degradation (e.g., depolymerization) than thermoplastics. Thus, it is also necessary to specify that thermoplastic materials contain heat stabilizers (to retard oxidative degradation of the composite) when designing with these materials for high temperature end uses.

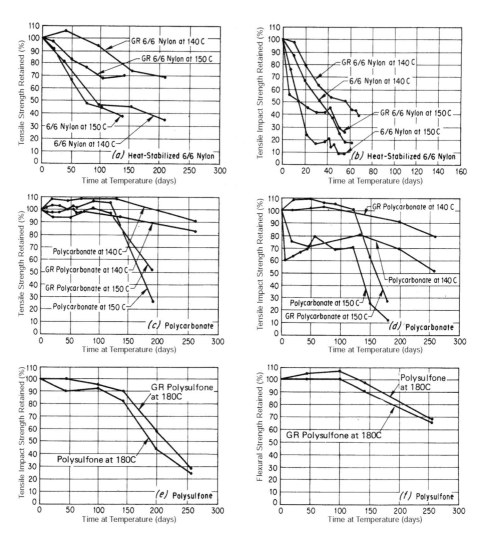

Figure 4.3 Effects of time and temperature on mechanical properties of nylon 6/6 (heat-stabilized), polycarbonate, and polysulfone in unreinforced and in 30% glass-reinforced (GR) compounds
(Courtesy LNP Engineering Plastics, Inc.)

4.6 Toughness

Toughness, or impact strength, has long been an enormous challenge for design engineers. The nature of the applied stress, the geometry of the part, and the characteristics of the material used, all combine and interact to make this property one of the most difficult and complex aspects of part performance to analyze and predict. The presence or absence of mechanisms to dissipate dynamic stress, such as the ability of the part to bend, twist, or deform, also play an important role in determining the ability of the part to withstand impact, as well as the effect of repeated blows. As noted earlier, the effect of short glass reinforcement is to change the mode of impact failure for ductile base resins to brittle (generally also accompanied by a reduction in energy-required-to-break); conversely, brittle base resins gain toughness, as the effect of reinforcement increases the energy-required-to-break. The "long glass" compounds exhibit significantly increased notched Izod and falling dart test values, but some investigators have found that this does not translate into higher toughness in actual part service, in terms of energy-to-break, unless an impact modifier has been added to the composite [7].

Crystalline polymers generally show a greater loss in toughness and increase in notch sensitivity at low temperatures than do amorphous polymers [11]. For this reason, suppliers have brought out "toughened" grades of such materials as nylons, acetal, and polypropylene. In one example, the additions of acrylonitrile–butadiene–styrene (ABS) (an elastomeric polymer) and glass fiber reinforcement to nylon 66 as the matrix polymer were found to contribute to improving the composite's resistance to crack propagation and crack initiation, respectively [12]. Some thermosets that can be processed and used at temperatures under 204 to 232 °C (400 to 450 °F) may be able to utilize polyethylene terephthalate (PET) fibers to enhance toughness, but this is one property where thermosets do not often compare well with thermoplastics, within the definitions for materials used in this book.

Most suppliers provide notched Izod impact strength (ASTM D256) values for their materials. Despite its notorious dependency on the radius of the notch in the test specimen and low confidence level in discriminating between materials, notched Izod has at least some utility as a measurement of crack propagation where the part design unavoidably incorporates a "stress riser" or abrupt cross-section change that serves to concentrate stress in a small area. Where part walls are more uniform, unnotched Izod or Charpy values may be more useful, as are tensile impact data (D1822), which have been found to correlate closely with each other [13]. Another widely used toughness test is the Gardner falling dart drop impact (ASTM D3029). A more sophisticated test is the instrumented impact test (ASTM D3763), which measures energy-to-break over a range of loading speeds.

All of these tests have significant limitations when applying the values to design. In the real world, parts are subjected to impact at velocities, temperatures, and locations that may not be easily included in the testing parameters. In fact, one pair of investigators [14]. found no statistical confidence in the rankings of materials tested by falling weight, notched Izod, or fracture mechanics methods, plainly showing that differing test methods measure differing aspects of toughness. This study strongly underlines the necessity of using actual part impact tests to determine the response of a fabricated material to impact stresses, and not to rely on one or more traditional laboratory impact tests.

The best general rules of thumb concerning the toughness characteristics of short fiber reinforced plastics are as follows:

- The longer the glass length, the greater the resistance to cracking from being impacted.
- The higher the glass content, the higher the strength but the lower the elongation.
- Applications requiring exceptionally high impact resistance should utilize "toughened" (elastomer-modified) base resins or "tough" reinforcements, for example, PET.
- Thermoplastics normally show better toughness than thermosets.
- There is no substitute for prototyping and testing!

4.7 Environment

Environment must always be considered as yet another influence on whether or not a material will perform as anticipated in a given design. However, the word environment covers a number of external conditions, which we will consider individually.

4.7.1 Chemical Attack

Many applications involve exposure of the plastic part or assembly to various chemicals—even a business machine housing, which theoretically should be located in a sheltered environment, may have to contend with coffee spills or detergents! Even if the exposure does not involve direct immersion or hard contact, periodic contact or merely contact with vapor may be sufficient to cause cracking in parts that are either under stress or retain molded-in stress. Fortunately, reinforcement tends to improve the ability of many polymers to resist stress-cracking, particularly amorphous ones such as polycarbonate.

Reactive chemicals cause damage by breaking the polymer chain through such processes as oxidation or hydrolysis. By far the most likely symptom of chemical attack is solvation, however. It is uncommon for polymers to dissolve completely in a solvent, as salt would in water, but rather for the solvent to diffuse into the polymer, causing it to soften and swell. Thus, chemical attack can cause dimensional changes as well as loss of mechanical properties in a composite. Again, fiber reinforcement tends to enhance property retention of the base polymer upon exposure to chemical reagents, including water.

Some reagents can cause plastic parts to crack at points where internal stress are present, either from molded-in stress or external applied mechanical loads. Some polymers are more likely to stress-crack than others: polyethylene and polycarbonate are well-known examples of polymers with this weakness. Although the phenomenon of stress-cracking has been studied for over 40 years, the mechanism by which it takes place has not yet been sufficiently characterized as to permit one to predict when it will occur. Fiber reinforcement as well as formulation with an elastomer can diminish this tendency but not eliminate it. Prototype testing is essential if a part will be under continuous load in contact with one or more chemical reagents.

Unfortunately, to the author's knowledge, there are no really useful data available that compare the chemical resistance of most plastic materials under uniform test conditions. Worse, there is no generally accepted simple method of predicting the performance of plastics materials exposed to various chemicals. Consequently, one is forced to make a crude screening of materials from the qualitative chemical resistance ratings furnished by materials suppliers (see Table 4.2) and then conduct some form of test on either sample specimens or a prototype part.

4.7.2 Dielectrics and Flammability

One of the growing uses for short fiber reinforced plastics is in electrical and electronic components where the ever-increasing miniaturization of devices and assemblies has resulted in higher operating temperatures. In addition, some assembly processes, such as vapor phase soldering of printed circuit boards, test the temperature limits of conventional materials. Plastics are used because they insulate against unwanted electrical current leakage (although, in some special cases, they can be formulated to actually conduct

Table 4.2 Chemical resistance of glass fiber reinforced thermoplastics; loss in tensile strength as indicated. E = excellent (0 to 3%), A = acceptable (3 to 10%), F = fair (10 to 25%), X = Unacceptable (<25%); **bold face** at 0.25% flexural strain

Polymer	ETFE	FEP	Nylon 66	PBT	PPS	PES	PSF
Glass Content	20	20	50	40	40	40	40
7 days at 23°C							
HC1, 10%	E,**E**	E,**E**	F,**X**	A,**A**	E,**E**	A,**A**	A,**F**
H_2SO_4, 10%	F,**F**	A,**F**	F,**F**	F,**X**	F,**F**	A,**F**	A,**F**
Water	E,**E**	E,**E**	F,**F**	F,**F**	E,**E**	E,**E**	E,**A**
NH_4OH, 10%	E,**E**	E,**E**	F,**X**	X,**X**	E,**E**	A,**F**	F,**F**
Ethylene Glycol	E,**E**	E,**E**	A,**F**	A,**A**	A,**A**	E,**E**	E,**E**
Toluene	F,**F**	E,**E**	E,**E**	F,**F**	A,**F**	F,**X**	X,**X**
Trichloroethylene	E,**E**	E,**E**	E,**A**	F,**F**	F,**X**	F,**F**	X,**X**
3 days at 82°C							
HC1, 10%	E,**E**	A,**A**	X,**X**	X,**X**	A,**F**	A,**F**	F,**F**
H_2SO_4, 10%	E,**E**	F,**F**	X,**X**	X,**X**	X,**X**	A,**F**	F,**F**
Water	E,**E**	E,**E**	X,**X**	X,**X**	A,**F**	A,**F**	F,**F**
NH_4OH, 10%	E,**E**	E,**E**	X,**X**	X,**X**	E,**A**	F,**F**	F,**F**
Ethylene Glycol	E,**E**	F,**F**	X,**X**	X,**X**	A,**F**	E,**A**	A,**F**
Toluene	F,**F**	F,**F**	A,**A**	X,**X**	F,**F**	F,**X**	X,**X**
Trichloroethylene	F,**F**	F,**X**	A,**F**	X,**X**	X,**X**	X,**X**	X,**X**
1 day at 149°C							
HC1, 10%	F,**F**	F,**F**	X,**X**	X,**X**	X,**F**	X,**X**	X,**X**
H_2SO_4, 10%	X,**X**	X,**X**	X,**X**	X,**X**	X,**X**	X,**X**	X,**X**
Water	F,**F**	F,**F**	X,**X**	X,**X**	X,**X**	F,**X**	X,**X**
NH_4OH, 10%	X,**X**	F,**F**	X,**X**	X,**X**	F,**X**	F,**X**	X,**X**
Ethylene Glycol	F,**X**	F,**F**	X,**X**	X,**X**	F,**X**	F,**X**	X,**X**
Toluene	F,**F**	F,**X**	F,**F**	X,**X**	X,**X**	X,**X**	X,**X**
Trichloroethylene	X,**X**	X,**X**	F,**F**	X,**X**	X,**X**	X,**X**	X,**X**

Source: Data courtesy of LNP Engineering Plastics, Inc.

currents). In the event that the part design requires that the material have direct contact with electrical currents, then that material also must not sustain combustion in the presence of a discontinuous source of ignition. While thermosets have long been the preferred materials for electrical and electronic applications, thermoplastics use is growing more rapidly. This is due to several factors, including the following:

- the growing availability of higher temperature resistant products;
- the need for improved toughness in consolidated parts (where the component and the housing are made as one piece); and
- the need for easily, economically fabricated, complex thin-walled parts in consolidated assemblies.

To predict the performance of a given composite, one must look specifically at the polymers, reinforcements, fillers, and additives used, because they will each contribute specific properties to the resulting composite. For example, carbon fibers can confer conductivity whereas glass fibers cannot; flame retardant additives may be needed to enhance the base polymer's resistance to ignition while certain polymers, such as polysulfone, do not require this addition because they are already inherently ignition resistant.

The effect of most short fiber reinforcements (exceptions are treated below) on the insulating and capacitive properties of base polymers is not of major significance compared to the enhancing effect on mechanical properties, particularly heat resistance. What effects there are, are generally positive, for example, lower dielectric constant, lower dissipation factor, higher dielectric strength and arc resistance. The effect on flammability is also mostly secondary; that is, the primary flame retardancy is conferred by the polymer itself or additives incorporated into the composite. The fibers are there largely to restore some of the strength and toughness lost as a result of the high concentration of nonreinforcing flame retardant additives.

Table 4.3 presents some selected materials and their dielectric/flammability properties. Note how much the UL TI differs from deflection temperature for individual composites—the difference between long- and short-term exposure. Oxygen index is included to demonstrate the difference between seemingly equivalent flame resistant materials—all test V-0 at the same thickness by UL Subject 94, but vary from only 27.2% O_2 required to sustain combustion to an impressive 50% or even more.

Perhaps one of the most intriguing effects of specific reinforcements on dielectric properties is the ability to provide electrostatic charge conductivity and shielding from electromagnetic (primarily radio) frequency interference. The addition of carbon fibers or metal fibers can do just this. Base

Table 4.3 Typical Dielectric/Flammability Properties of Glass Fiber Reinforced Plastics (All UL Subject 94 V-0 Rated at 3.2 mm Thickness)

	UL temperature index, °C (°F) (electrical)	Deflection temperature, °C @ 1.82 Mpa (°F @ 264 psi)	Dielectric strength (short time), KV/cm (volts/mil)	Dielectric constant @ 60Hz/10⁶Hz	Dissipation factor × 10⁻⁴ @ 60Hz/10⁶Hz	Arc resistance, seconds	Oxygen index % O₂
UL/ASTM	UL 746B	D648	D149	D150	D150	D495	D2863
Base resins: Thermoplastics (30% glass)							
ETFE	170(338)	238 (460)	162 (410)	3.5/3.4	6/50	75	>75
LCP	220(428)	310 (590)	201 (510)	4.6/ –	80/ –	190	35–50
PA 6	130(266)	207 (405)	177 (450)	4.3/3.8	100/175	110–120	28
6/6	130(266)	243 (470)	177 (450)	4.4/3.8	90/190	110–120	36
610	65(149)*	204 (400)	197 (500)	4.4/4.0	125/160	100	28
PC	125(257)	149 (300)	177 (450)	4.2/3.5	8/75	5–120	30–35
PPE/PS blend	100(212)	110 (230)	193 (490)	3.0/3.1	40/80	70–120	28
PPS	220(428)	260 (500)	197 (500)	3.8/3.7	20/40	10–125	40
PP	110(230)	143 (290)	197 (500)	2.9/2.8	40/140	15–40	27

PSO	160(320)	185 (365)	189 (480)	3.6/3.5	20/50	110	35–40
PTFE (25% glass)	180(356)	NA	508 (1290)	2.6/2.9	718/28	>300	>75
PVC (20% glass)	50(122)	93 (200)	158 (400)	3.0/3.8	90/170	60–80	42
SAN	90(194)	93 (200)	177 (450)	2.8/2.8	20/300	60–80	32
PBT	140(284)	204 (400)	189 (480)	3.7/3.6	20/180	80–130	34
PET	150(302)	224 (435)	169 (430)	3.7/3.6	35/130	80–130	42
Thermosets (40–50% glass)							
Alkyd	130(266)	232 (450)	375 (193)	5.5/5.0	20/200	180	—
Allyd (DAP)	130(266)	232 (450)	400 (206)	4.6/5.9	20/100	140–180	—
Epoxy	130(266)	260 (500)	350–400 (180–206)	4.5/5.0	100/200	130–180	—
Amino (Melamine)	100–150(212–302)	204 (400)	350 (180)	– /7.0	– /170	140–180	—
Phenolic	150(302)	191–288 (375–550)	300–425 (155–219)	6.0/5.0	90/300	180	—
Polyester	130(266)	249 (480)	37–425 (191–219)	4.6/4.9	85/300	130–170	—
Silicone	105(221)	260 (500)	350 (180)	– /4.0	– /30	230	—

* "Generic," not specifically tested

polymers normally have surface resistivities (ASTM D257) of 10^{15} to 10^{18} ohms/m². The addition of 5 to 30% of the above fibers can lower the surface resistivities of the matrix resin down to 10^6 to 10^2 ohms/m², which will provide for gradual dissipation of electrostatic charges, before they reach such levels as to generate a spark. This ability is particularly useful for the protection of sensitive computer components, such as integrated circuits, which can be damaged by uncontrolled electrostatic discharge. Electromagnetic frequency shielding requires still lower surface resistivity, below 10^2 down to 10^{-2} ohms/sq. These applications may require fiber loadings up to 50% to create the conductive pathways necessary for such low resistivity levels. Carbon fibers themselves have surface resistivities in the 10^{-1} to 10^{-3} ohm/sq. range; the resistivity of metals runs in the range of 10^{-5} to 10^{-6} ohms/sq. Table 4.4 provides a sampling of properties of some of these materials.

Of course, there are nonfibrous conductive additives such as carbon black (often used in plasticized PVC) or silver oxide coated glass spheres (used in silicones for gasketing). These products are seldom used in conjunction with fibrous reinforcements, however. Typically they are used in flexible, nonstructural applications and are, therefore, considered to be outside the scope of this book.

4.7.3 Wear

Another specialized and complex environment is wear—a general term that includes abrasion and scratching. Wear is of particular interest when one is designing gears, cams, ratchets, slides, bearings, and other moving parts. Wear takes place as surfaces in contact slide past each other. Wear is primarily a function of the different coefficients of friction and hardnesses of the surfaces, the load applied to them, and the speed at which they move past each other. Other factors, such as temperature and chemical environment, also can have a significant effect on wear (once again, the need for prototyping and testing is obvious).

The use of fibers in a composite increases surface hardness which typically improves the wear resistance over that of the base polymer used in the matrix. Glass fibers also increase the wear on the mating surface and will increase the coefficient of friction of the composite unless combined with a lubricant (see Table 4.5). Carbon and aramid fibers both improve the surface hardness and other physical properties of the matrix, and reduce the coefficient of friction, although less efficiently than such additives as silicone oil or PTFE powder. Aramid fibers also cause markedly less wear on metal mating

Table 4.4 Typical Properties of Conductive Fiber Reinforced Thermoplastics

Base polymer	Reinforcement, wt. %	Specific gravity	Tensile modulus, GPa (Mpsi)	Notched Izod impact strength, J/m (6.4 mm bar) (ft-lb/in. (¼" bar))	Surface resistivity, ohms/sq.	Shielding effectiveness db attenuation @ 10^3 MHz, 6.4 mm (⅛") thick.
ASTM		D792	D638	D256	D257	ES7
PC	5% Stainless steel	1.28	4.83 (0.70)	80.1 (1.5)	10^2	40
PA 6/6	5% Stainless steel	1.22	4.41 (0.64)	42.7 (0.8)	10^2	40
PPS	15% Carbon fiber nickel, coated	1.45	6.89 (1.00)	32.0 (0.6)	10^3	20
PA 6/6	15% Carbon fiber nickel, coated	1.20	7.58 (1.10)	37.4 (0.7)	10^1	55
PPS	40% Carbon fiber	1.49	30.30 (4.40)	80.1 (1.5)	10^2	30
PA 6/6	50% Carbon fiber	1.38	34.5 (5.00)	106 (2.0)	10^1	50
PA 6/6	40% Carbon fiber	1.33	29.3 (4.10)	118 (2.2)	10^2	40
PA 6/6	30% Carbon fiber	1.28	20.7 (3.00)	107 (2.0)	10^2	30
PC	30% Carbon Fiber	1.38	15.9 (2.30)	96.1 (1.8)	10^2	40

Table 4.5 Typical Properties of Wear-Resistant Short Fiber Reinforced Plastics

| Base polymer | Reinforcement, lubricant, wt. % | Wear factor @ 23°C, 10^{-8} cm³·min./m/kg/h (@ 73°F, 10^{-10} in.³·min/ft/lb/hr) | Coefficient of friction | | Limiting PV (Kpa-m/s × 10^{-3}) @ 23°C and 30.5 ml/min ((psi-fpm) @ 73°F and 100 fpm) | Deflection temperature °C @ 1.82 Mpa (°F @ 264 psi) | Water absorption % in 24 h |
			Static @ 23°C (276 KPa) (@ 73°F (40 psi, 50 fpm))	Dynamic @ 23°C (276 KPa, 15.2 ml/min) ((psi-fpm) @ 73°F and 100 fpm)			
POM	None	77 (65)	0.14	0.21	51 (3,500)	110 (230)	0.22
	30% Glass fiber	290 (245)	0.25	0.34	—	163 (325)	0.60
	30% Glass fiber + 15% PTFE	237 (200)	0.2	0.28	256 (17,500)	160 (320)	0.27
	20% Carbon Fiber	47 (40)	0.11	0.14	292 (20,000)	160 (320)	0.50
Nylon 6/6	None	237 (200)	0.20	0.28	37 (2,500)	104 (220)	1.50
	30% Glass fiber	89 (75)	0.25	0.31	146 (10,000)	254 (490)	0.09
	30% Glass fiber + 15% PTFE	19 (16)	0.19	0.26	292 (20,000)	254 (495)	0.50
	30% Glass fiber + 15% PTFE + 2% Silicone	11 (9)	0.12	0.14	292 (20,000)	257 (495)	0.45
	30% Carbon Fiber	24 (20)	0.16	0.20	394 (27,000)	254 (490)	0.50
	50% Glass fiber + 5% MoS₂	89 (75)	0.24	0.31	219 (15,000)	254 (490)	0.80
	20% Aramid Fiber	73 (62)	0.22	0.25	—	249 (480)	0.65

PPS	None	640 (540)	0.30	0.24	44 (3,000)	138 (280)	0.05
	40% Glass fiber	28 (24)	0.38	0.29	234 (16,000)	263 (505)	0.02
	30% Glass fiber + 30% PTFE	89 (75)	0.13	0.15	511 (35,000)	260 (500)	0.03
	20% Carbon Fiber	190 (160)	0.23	0.20	292 (20,000)	263 (505)	0.04
ETFE	None	5,925 (5,000)	0.05	0.40	—	71 (160)	0.02
	30% Glass fiber	12 (10)	0.17	0.18	—	238 (460)	0.02
	30% Carbon Fiber	7 (6)	0.11	0.20	—	241 (465)	0.02
TPU	None	403 (340)	0.32	0.37	22 (1,500)	32 (90)	0.40
	30% Glass fiber	213 (180)	0.30	0.34	—	171 (340)	0.25
	30% Glass fiber + 15% PTFE	41 (35)	0.20	0.25	146 (10,000)	85 (185)	0.35
Phenolic	40% α-Cellulose Fiber + 10% PTFE	356 (300)	—	(0.16)	511 (35,000)	163 (325)	0.70

Source: LNP Engineering Plastics, Inc.

surfaces. The addition of elastomers can also help improve wear resistance of the composite.

Fiber orientation is important in designing for minimum wear. Ideally, one should design so that wear takes place on a surface where the fiber ends are perpendicular to the surface, so that the fibers are deeply embedded in the matrix and least likely to be torn out. The next best orientation would be with the sliding motion taking place at right angles to the fiber length; the least desirable is for the sliding motion to be along the axis of the fiber, as the fiber is least strongly anchored to the matrix in this situation. Table 4.5 presents data on selected wear-resistant short fiber reinforced composites.

4.7.4   Radiation

Radiation includes exposure of the composite to sunlight (mainly ultraviolet rays) and irradiation by α, β, and γ rays. Consideration of materials functioning in these environments is really limited to the performance of the base polymers, as the reinforcements have no specific direct contribution to the resistance of the composite to degradation, with one exception. The latter is carbon fiber, which inhibits degradation from ultraviolet radiation by absorption, as carbon black pigmentation at a level of 2.5% or more will achieve much the same result.

Inexplicably, acrylic polymers, which offer the best UV resistance of any thermoplastic, are not at present available as reinforced composites on a commercial basis. Perhaps this is a market niche that will be filled in the future. At least for now, designers must fall back on UV-stabilized and/or highly pigmented (the darker the better) polymers for the matrix.

Resistance to nuclear radiation is a very specialized requirement. Alpha radiation is seldom encountered except in the direct presence of radioactive materials; it is nonpenetrating and therefore of minimal effect on plastic materials. Beta and gamma radiation is encountered in food and medical sterilization, as well as in more exotic settings such as nuclear/fusion power, particle physics research, or even spacecraft and satellites. As a general rule, the more highly aromatic the molecules in the chemical composition of the polymer, the more resistant they are to radiation. Therefore, such polymers as polyimide, epoxy, polysulfone, and polyphenylene sulfide will outperform polyesters (both thermoset and thermoplastic), acetal, polyolefins, and polyvinyl chloride.

4.8   Fabrication

As noted earlier, the peculiarities of fabrication must ever be kept in mind when designing a part. Often, these present limits on what can be done, but they can also present opportunities to enhance the performance of the part, such as when the orientation of fibers in long flow paths can be used to strengthen a part against stress in the flow direction. Tensile, flexural, and shear strength properties, however, which are generally reported based on measurements in the flow direction, are likely to be as much as 40% lower when measured at right angles to the flow direction. Compression strength is not as greatly affected. These differences are most pronounced in designs that have relatively thin sections (so that the highly oriented skin makes up the greater part of the overall wall thickness) and, of course, in sections with long flow paths. While a "thin" wall has been defined typically as one of 3.8 mm (0.15 in.) or less, wall thickness as small as 0.26 mm (0.030 in.) are now being designed in cellular phone and laptop computer applications [15]. A "long flow path" is considered to be one where the length of flow is roughly ten or more times the width of the path (with a "thin" wall). In the past few years, suppliers have recognized the need to formulate materials that offer high flow combined with high strength and modulus, allowing engineers to design parts that have thinner walls than could be designed in older materials. The potential weight reduction may make a higher specific cost compound having high property values more attractive than a lower specific cost compound having lower property values. Applications where the wall geometry is constrained are also often ideal for "long glass" compounds.

Other concerns include minimizing weld lines and sink marks, avoiding fiber length attrition, and attaining either high or low surface gloss. These factors are usually controllable by mold design and processing conditions. The location of parting lines and cores can significantly affect the performance of the part and the cost of tooling and fabrication. These all require close and timely consultation between the design engineer, the mold builder, and the molder as the work of each will critically affect the quality, cost, and acceptability of the finished part. More detail on this subject is found in Chapters 7 and 8.

4.9   Economics

A rough rule of thumb commonly used in the past by custom molders for estimating the cost of a part is to calculate the material cost and then double it. Although this has a certain attractive simplicity, it takes but a moment's thought to see how it may be easily misused and liable to lead to wildly erroneous estimates. For example, a part may be designed in glass fiber reinforced nylon 6/6, and then changed to carbon fiber reinforced nylon 6/6. Although the material cost might rise by a factor of four, the fabrication cost would be essentially unchanged. Obviously, doubling material costs to get ballpark estimates of part costs simply does not work well for components designed in the more expensive engineering plastics.

While it is true that price per unit of weight is only one aspect of the cost of a fabricated part, nevertheless, it is often the most important and, therefore, the first one to be estimated. The designer must remember that materials do vary considerably in density, and that part costs must be based on the volume cost of the material to be used, not merely its price per kg or Figure 4.4 shows a nomograph that can assist with this step.

Mentioned earlier, the use of computer aided design programs, such as Design for Assembly and Manufacture (DFMA), enable the design engineer to have the greatest impact on product costs before they become largely unchangeable. Boothroyd and Dewhurst note that "70% of all product development, assembly and manufacturing costs are built-in during the design stage" [16]. Material selection is not enough, even though critical. The designer must also compare, analyze, and choose the simplest, most cost-effective product based on the number of parts, fastening, and assembly techniques.

Fewer components in an assembly mean savings in labor, inventory, and capital investment. Fewer components also mean greater control over quality, and more efficient marketing and distribution through reduced returns, spare parts, and service. Fewer components mean reduced administration costs in purchasing, accounting, and floor space. Simpler products mean shorter product-to-market cycles, providing a competitive edge. Short fiber reinforced plastics are an important part of this process, because they perform more reliably than unreinforced materials while still retaining the most basic advantage of plastics: ease of low-cost fabrication.

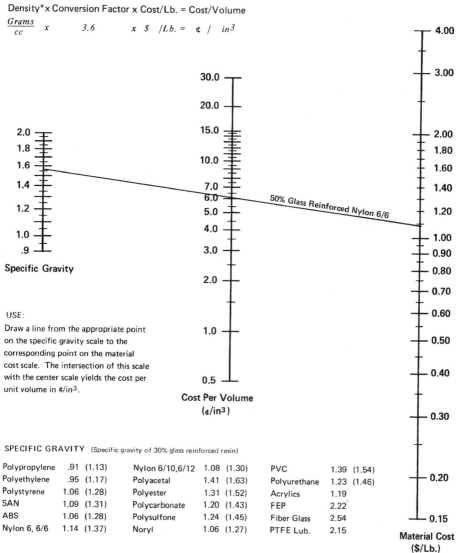

CALCULATION:
Density* x Conversion Factor x Cost/Lb. = Cost/Volume

$$\frac{Grams}{cc} \quad x \quad 3.6 \quad x \ \$ \ /Lb. = \ ¢ \ / \ in^3$$

4.00

3.00

30.0

20.0

15.0

10.0

2.00

1.80

1.60

2.0
1.8
1.6

7.0
6.0
5.0

1.40

50% Glass Reinforced Nylon 6/6

1.4

1.2

1.20

4.0

1.00

1.0
.9

3.0

0.90

Specific Gravity

0.80

2.0

0.70

0.60

USE:

Draw a line from the appropriate point
on the specific gravity scale to the
corresponding point on the material
cost scale. The intersection of this scale
with the center scale yields the cost per
unit volume in ¢/in³.

1.0

0.50

0.40

0.5

**Cost Per Volume
(¢/in³)**

0.30

SPECIFIC GRAVITY (Specific gravity of 30% glass reinforced resin)

Polypropylene	.91 (1.13)	Nylon 6/10,6/12	1.08 (1.30)	PVC	1.39 (1.54)
Polyethylene	.95 (1.17)	Polyacetal	1.41 (1.63)	Polyurethane	1.23 (1.46)
Polystyrene	1.06 (1.28)	Polyester	1.31 (1.52)	Acrylics	1.19
SAN	1.09 (1.31)	Polycarbonate	1.20 (1.43)	FEP	2.22
ABS	1.06 (1.28)	Polysulfone	1.24 (1.45)	Fiber Glass	2.54
Nylon 6, 6/6	1.14 (1.37)	Noryl	1.06 (1.27)	PTFE Lub.	2.15

0.20

0.15

**Material Cost
($/Lb.)**

*The data sheet specific gravity
can be used as the density
value in this calculation.

LIQUID NITROGEN PROCESSING CORPORATION
PENNSYLVANIA
412 King Street, Malvern 19355/215-644-5200
CALIFORNIA
1831 E. Carnegie, Santa Ana 92705/714-546-2000
EUROPE
LNP Plastics Nederland B.V./Kalshoven, 11/Breda
The Netherlands/076-873150

ENGINEERING PLASTICS

© 1975

PRINTED IN U.S.A.

Figure 4.4 Cost per cubic inch comparator
(Courtesy of LNP Engineering Plastics, Inc.)

4.10   Ecological Concerns

Finally, the designer cannot ignore the fact that his or her product will be eventually disposed of, and that such disposal should be ecologically sound. Fortunately, short fiber reinforced plastics offer no threat to the environment in themselves—they are not toxic or inherently dangerous to health and safety. The question to be addressed is more one of what is to be done with the items made from these materials when they have reached the end of their service life. There are three commercial options at present for the disposal of products made from plastics materials:

- incineration,
- recycling,
- landfill.

Because this field is still in development, no one of these methods has yet been established as "the" one, and, in fact, it is a virtual certainty that each will have its role in the overall solution. Nevertheless, there are some pragmatic considerations that should be noted as particularly applicable to short fiber reinforced plastics.

First, the leading reason why these materials have become so important is their durability. Therefore, it would be illogical to incorporate an agent that would permit biodegradation. This remedy may have some merit for packaging materials that could wind up as roadside litter, but short fiber reinforced plastics are too costly to be used as disposable packaging materials. Their durability would make them suitable for landfill. Their suitability derives from their inertness: they will not dissolve or crumble, making the landfill unstable or a threat to contaminate groundwater.

Second, modern, high-temperature incineration would certainly be an acceptable method of disposing of these materials. Although some ash would remain, it would be primarily nontoxic glass or mineral. Short fiber reinforced plastics should present no unusual problems for disposal by this method, which is widely employed in Europe.

Last, recycling offers an ecologically sound method of disposal for short fiber reinforced plastics. Standards now in use require the molded-in labeling of the material used in a plastic part, so as to facilitate the separation of materials for reuse. However, the label now used for engineering plastics is "Other," which provides minimal information and prevents primary sorting of collected products using simple inspection. Further refinement of the

"Other" category would seem worthwhile, but will require discussion among industry, technical, and government personnel to implement recycling on a widespread scale [17]. Considering the relatively high costs of short fiber reinforced thermoplastics, reuse economics are potentially favorable. Recycling economics are very nearly the same for thermoset materials (the recyclate can be used as an essentially inert filler).

At present, there is some movement to recycle engineering plastics, both reinforced and neat materials: several polymer producers and compounders have announced programs to take back molding scrap and rework this material for reuse by the processor for a recycling fee. An advantage to this procedure is that the recycled material has a documented history and is likely to be relatively free of contaminants. Further, it can be tested and certified to specific properties. These are "post-industrial" rather than "post-consumer" recyclates, it is true, but it looks like it may be a long time before engineering plastics will be collected after consumer use in any significant volume. Economics, however, will always be the key to successful recycling programs.

An interesting illustration of the recycling potential for mixed fiber reinforced plastics has been reported by Hoechst Celanese Corporation, producers of acetal resins and compounds. They found that glass fiber reinforced acetal copolymer could be blended with polypropylene in concentrations as high as 15%, with minimal loss of physical properties of the polypropylene. This could be very useful in the recycling of automobile interior components, which currently contain an average ratio of 1 part acetal to 17 parts polypropylene (or 6%) [18].

References

1. Engelstein, G. "Integrating Plastic Process Simulation into the Corporate Structure: Marketing and Management Implications," Society of Plastics Engineers Annual Technical Conference, San Francisco, CA, May 1994
2. Anon. "Where Process Simulation *Doesn't* Work," *Inject. Mold.* November 1994
3. Holtz, W.B. *The CAD Rating Guide,* 5th edit. (1998) Penwell Publishing, Westlake Village, CA
4. Miller, B. "Predicting Part Shrinkage Is a Three Way Street," *Plastics World,* December 1989
5. Rosato, D.V. "Injection Molding Higher Performance Reinforced Thermoplastic Composites," Society of Plastics Engineers Annual Technical Conference, Indianapolis, IN, May 1996
6. Klein, A. "Composites That Fight Fatigue," *Adv. Mater. Proc.* February 1986

7. Kim, H.C., Glenn, L.W., Ellis, C.S., Miller, D.E. "Creep Behavior of Long and Short Glass Fiber Reinforced Thermoplastics," Society of Plastics Engineers Annual Technical Conference, Indianapolis, IN, May 1996

8. Sepe, M.P. "The Use of Advanced Characterization Techniques in Evaluating the Fitness for Use of Long Glass Fiber Thermoplastics," Society of Plastics Engineers Annual Technical Conference, San Francisco, CA, May 1994

9. Sepe, M.P. "A Practical Approach to the Glass Transition for Design Engineers," Society of Plastics Engineers Annual Technical Conference, Indianapolis, IN, May 1996

10. Sepe, M.P. "Proposed Enhancements to the Short Term Property Chart for Improved Material Decisions," *ibid*

11. Wolverton, M.P., Theberge, J.E. "Reinforced Thermoplastics That Fight The Cold," *Machine Design* (1979) XX, April 12

12. Wong, S., et al. "Toughening of Nylon 6,6/ABS Alloys," *Plast. Eng.* January 1995

13. Theberge, J.E. "Impact-testing of Glass Fortified Thermoplastics," *Mod. Plast.* (1969) 46, July

14. Friedrich, K., Walter, R. "Fracture Toughness of Short Fiber/Thermoplastic Matrix Composites, Compared on the Basis of Different Test Methods," Proceedings of the American Society for Composites Second Technical Conference, Newark, DE, Sept. 1987

15. Miller, B. "Ready for the New Wave of Super-Thinwall Parts?, *Plastics World,* May 1996

16. Boothroyd, G., Dewhurst, P. *Concepts for the Future of Manufacturing* (1987) Boothroyd Dewhurst, Inc., Wakefield, RI

17. Jones, R.F., Baumann, M.H., "Recycling of Engineering Plastics Advances," *Mod. Plast.,* May 1998

18. Naitove, M. "PP/Acetal Blends are Recyclable," *Plast. Technol.* March 1996

5   Prototyping and Testing

Mitchell R. Jones

5.1   The Need for Prototyping

Physical prototyping can save time and money. The process of manufacturing a prototype part and testing it under simulated end-use conditions increases the likelihood that the part will meet customer quality requirements, improves time to market, and minimizes the risk to investments in production tooling. The goal is to expose and correct functional shortcomings of the design, evaluating part geometry, material, and fabrication method (including mold flow effects). Prototyping can determine something as simple as whether the entire mold can be easily filled or something as fundamental as whether the material is capable of meeting the application requirements.

Prototyping is a worthwhile investment for all but the most elementary of designs. To some degree, computer simulation techniques provide reliable prediction of fiber orientation, polymer orientation, and weld line location—the principal properties in short fiber reinforced plastics that are affected by fabrication variables. Finite element analysis (FEA) also offers some guidance to design optimization before prototyping. Nevertheless, until these simulation techniques are proven accurate and reproducible for the design in question, judgment and prototyping will be the primary tools for optimizing a design.

It may be necessary to repeat prototype part production, testing, and design modification several times to arrive at the optimum material and geometry. This is needed because judgment is often incomplete and sometimes incorrect, so that the initial design may produce a part with inadequate properties. In addition, each prototype phase may have a different purpose and require a different approach. In many cases, it is more effective to examine specific property aspects of a component in the early prototype phases, then simulate actual service later in the evolution of the design.

The quickest and most economic prototyping process depends on a number of factors, primarily the following:

- the experience of the design engineer,
- the sophistication and precision of the design tools available and the complexity of the design,

- the number of parts to be produced and size of the investment in production of materials and tooling, once the design is finalized,
- the potential cost of a failure to the end user, and
- the intended service life and durability of the component.

These factors need to be considered when choosing which properties to examine most carefully and the required confidence level.

5.2   Prototype Techniques

There several prototype techniques available. First, one must decide whether to make a model of the part or to skip this stage and to attempt actual prototype production with the originally designed geometry and material. A model part may use a material that behaves like the design material, but does not necessarily have the same exact dimensions as the final design. Next, the part is subjected to an environment, for example, load, heat, or chemical attack, that simulates some or all aspects of the application.

The choice of prototype technique is best determined by assessing what properties of the component are unknown at the time. What is of greatest interest: stiffness, yield load, or heat distortion temperature? This will help to determine the simplest and least expensive prototyping technique.

Techniques for creation of a physical prototype component include the following:

- injection or compression molding with a low-cost mold,
- machining from a block of similar material,
- assembly, for example, by adhesive bonding, of standard shapes,
- hand-layup of a "model" material, and
- thermoforming of sheet and other standard shapes.

Short fiber reinforced plastics are not homogeneous (properties vary from local point to point) nor are they isotropic (properties vary with direction of measurement). Some of the above methods will not reproduce the flow patterns that occur during production molding, so they cannot be used to study properties that are dependent on fiber orientation, such as modulus, yield stress, or stresses induced by shrinkage. In addition, prototypes made from slab or bar stock are based on extrusion grades of the polymers involved, which usually have higher molecular weight than injection molding grades. This, in turn, will often cause the prototype to exhibit better impact strength, creep resistance, and chemical resistance than the molded part. These differences are particularly pronounced with crystalline polymers such as nylon, acetal, and polypropylene. Even the smaller differences observed with amorphous polymers such as polycarbonate and modified polyphenylene ether (PPE) may be critical to a given application. Machining removes the "skin" (often resin-rich) from prototypes parts, introducing further deviation from the production versions.

Consequently, injection molding with a low-cost mold is the most reliable technique in designs based on short fiber reinforced plastics.

Note, however, that prototype molds still introduce another variable: cooling rate. The less similar the cooling rate of the material inside the prototype mold vs. inside the production mold, the less realistic the fiber orientation-dominated properties of the prototype part will be. This is due to the dependence of skin and core flow effects on cooling rate. The two different types of flow produce markedly different fiber orientations, and their relative thicknesses strongly affect the in-plane properties of molded components. Therefore, the thermal conductivity and the heat capacity of material used for the prototype mold should approximate that of the material used for the production mold (which is usually steel).

The most economic compromise for a prototype mold material is often aluminum or epoxy filled with aluminum powder. The latter material has a thermal conductivity approaching that of solid aluminum, which is nearly five times that of mold grade steel (and therefore introduces a variant into the prototype process). Either material can be machined into the correct cavity shape; filled epoxy has the additional capability to be cast around a part form.

5.3   Prototype Testing and Evaluation

Once the physical prototype is created, it can be subjected to tests and evaluated under some or all aspects of the application environment. Mechanical, chemical, microstructural, electrical, visual, frictional, and combustive properties are usually relevant in some combination (see Table 5.1 for a checklist). Testing may also be divided into two categories: simulation testing in the laboratory, and service testing where the prototype is placed in actual use.

Laboratory testing is more economical where environmental conditions are not complicated by coupling (e.g., high moisture and temperature-accompanying mechanical stress) in the application. Laboratory testing also

Table 5.1 Preferred Nondestructive Evaluation Methods versus Types of Defects

Method	Cracks	Weld lines	Porosity	Fiber concentration	Fiber orientation
Visual	—	+	—	x	x
Microscopy	+	++	+	++ (requires destructive sectioning)	++ (requires destructive sectioning)
X-Ray	+	+	+	++	++
Ultrasonic	+	++	++	—	+
Thermograpy	—	—	+	+	x
Fluorescent dye	++	++	++ (surface only)	x	x

Key: ++ = effective; + = sometimes effective; — = ineffective; x = not applicable

allows isolation of the individual parameters, a useful feature if the application involves a complex and varied environment or if the behavior of the prototype is not predictable. Accelerated simulation of in-service conditions is often feasible by applying environmental factors that are most influential in component life time, such as stresses, strains, temperature, and humidity fluctuations; sunlight exposure; and attack by solvents. Several years of service can be simulated in the laboratory in just a few days but requires close attention to validity. For example, mechanical stresses can be applied in a shorter amount of time or at higher values than will occur in service, but only if such foreshortening does not invalidate the test. Applying stresses at a greater rate is valid if the increased rate will not cause a different response in the material due to effects such as self-heating or reduced severity of chemical attack. Using higher stresses than the real application may cause creep, crack initiation, or fracture that would not occur at lower stresses over a longer time period and must therefore usually not be attempted. Temperature or humidity fluctuations can be accelerated only to the point of maintaining uniform penetration that is likely in the end use enviroment. If creep or vibration is expected in service, "time-temperature superposition" may often be applied to accelerate laboratory testing. This technique mathematically predicts the material's response in service, based on laboratory characerization of the material over a range of temperatures (but at low strains). The prototype can then be tested at a lower temperature or rate of stress than will occur in service, and the effect of the same stress applied over a much longer time or the effect of vibration at a higher frequency may be predicted. Figure 5.1 shows a typical mechanical properties testing system.

Figure 5.1 A universal testing system for mechanical testing of components and standard specimens. These systems are capable of precise force or displacement controlled tests under monotonic or fatigue conditions. Such testing systems are widely used for quality control of production pieces as well as prototype evaluation and failure analysis.
(Courtesy of Instrom Corp.)

Service testing (or "field trial") includes all aspects of the application environment, notably those that may be unforeseen by the design engineer, and other variables such as the compounding effects of stress and temperature. For example, vibration may cause self-heating (resulting in altered mechanical properties) that was never anticipated by the designer. These same factors also complicate failure analysis, sometimes to the point that laboratory testing is also needed to establish the cause of part failure. If service testing will require weeks (or months) of data gathering, it is best reserved for cases where environmental conditions are unpredictable or failure analysis should be uncomplicated.

These techniques are also useful for production "proving" during the first runs and also for failure analysis if the production design fails to perform in the application.

The following sections review laboratory Nondestructive Evaluation (NDE), microscopy, and experimental stress analysis methods, which the engineer can use to obtain information about the presence and severity of flaws developed by prototype or test parts when they are subjected to sustained mechanical loading.

5.3.1 Nondestructive Evaluation (NDE)

NDE methods have the advantage that they cause no harm to the specimen; thus the same part can be nondestructively retested or subsequently tested destructively. However, NDE only reveals the location and severity of flaws. The experimenter must judge the importance of each particular flaw. Sometimes, the importance of a defect is obvious (harmless, or likely to foreshorten service lifetime, or likely to cause catastrophic failure); otherwise, experimental stress analysis will quantify the severity of the flaw. NDE methods include:

- visual inspection,
- microscopy,
- X-radiography using metal-coated fibers or tracers,
- ultrasonic wave propagation,
- thermography, and
- crack detection methods, such as fluorescent dye penetration.

The defects that are relevant to the strength of short fiber composites are as follows:

- surface cracks (due to excessive rate of cooling, for example),
- weld lines (where two polymer flow fronts converge while molding),
- areas of undesirable fiber concentration (due to undispersed fiber clumps or resin-rich areas, for example),
- areas of undesirable fiber orientation (such as predominant orientation transverse to the direction of maximum stress, or random orientation where alignment is preferred), and
- excessive porosity or visible voids (gas bubbles).

Table 5.1 illustrates which method is effective for a particular type of defect.

5.3.1 Micrography

It is usually appropriate to begin any prototype evaluation with a microscopic examination (micrography) of the part's structure: at a failure point, at the surface, and of several key cross-sections. Using typical magnification scales of 40× to 200×, sliced and carefully polished sections will reveal fiber orientation, porosity, cracking, and crazing. Thus the experimenter may evaluate a prototype design after fabrication, but before physical testing (to detect any obvious problem areas), as well as after, as an aid in failure analysis.

In some design situations, the locations of critical areas of stress are obvious. Proper mold design and choice of part geometry should ensure adequate strength in these areas, but microscopic observation of the fiber orientation and polymer flows is a prudent precaution. Micrography is typically cheaper than physical testing, often the most cost-effective way to determine the next step early in a prototype program.

Microscopic examination is more often beneficial when examining a part after physical testing. For example, a region that suffered unexpected cracking or complete loss of mechanical integrity may show inappropriate fiber orientation or voids formed during molding. As another example, chemical attack may cause damage in the form of matrix crazing. This insidious damage is most easily detected with micrography.

One should examine from one to ten or more locations in a prototype part. Each location may require one or more sections. Each section is obtained by cutting a small block of material from the part. This is usually accomplished with a carbide or diamond saw. The specimen is then embedded in a cylinder of resin for handling convenience. Working from coarse to fine abrasive, the specimen is then polished until its appearance is adequately smooth under microscopic examination. The smallest remaining scratches should be at least five times smaller than the smallest feature of interest. This will ensure that details, such as crazing, are not obscured or go undetected.

Approximate specimen preparation equipment costs range from $2000 to $20,000, while microscope costs are anywhere from $2000 to $40,000. The higher costs are associated with the ability to observe increasingly fine details, stereo viewing, photographic capability, etc.

5.3.3　Experimental Stress Analysis

The greatest uncertainty of a design is often whether it will withstand mechanical loads and displacements. Examining critical areas of high stress or strain is most precisely quantified using experimental stress analysis methods.

All but the most complex experimental stress analysis techniques are limited to measurement of strain on the surface of the prototype. The surface is often where stresses are greatest, for example when twisting or bending are present, so most analysis techniques apply. (When required, through-the-thickness data may be obtained by "slicing" the prototype and analyzing the slices separately.) The analysis methods are divided into "local" and "full-field" categories. A local analysis reveals strain only at a point, while a full-field analysis reveals strain at all points (a strain field). If the important localized stress points are not easily identified, then a technique that gives a full strain field is far more useful. Precise full-field methods are generally more expensive and time consuming.

Common techniques for experimental stress analysis are as follows, in decending order of the resolution and quantity of data they yield:

- **Full-field**
    Holographic interferometry
    Speckle interferometry (full-field type)
    Photoelastic stress analysis
    Moire interferometry
    Crackle lacquer coating
- **Local**
    Adhesively bonded strain gages
    Speckle interferometry (single or dual spot type)
    Optical extensometry (video or laser-based)

An inexpensive initial evaluation may be accomplished by placing a crackle-laquer-coated part under typical environmental loads. After observing the result, the engineer can select from among the above techniques to develop more information.

Holographic interferometry measures displacements optically by reflecting light (often a laser source) first off an undeformed prototype and then off a mechanically deformed one. By superimposing the images as a holograph, lines of equal displacement appear on the prototype because of interference between laser light reflected by the separate images. The result is a "map" of the displacement field where displacement magnitude and direction are known at every point. Areas of high stress concentration are highlighted by closely spaced interference lines. The displacements can subsequently be transformed into strains and then stresses by computer software. The holograph may be recorded on film, video tape, or a computer. This may be done several times by loading the prototype incrementally. Holography can accurately and quantitatively measure static and dynamic displacements. Sensitivity is very high, but measurements are time consuming and expensive.

Full-field speckle interferometry is similar to holographic interferometry except that the speckles are observed on the specimen suface by illuminating it with laser light (Fig. 5.2) or by establishing an optically speckled surface (for example by painting it), as in Fig. 5.3.

Photoelastic stress analysis is a powerful full-field technique where a prototype part is first covered with a thin film of special, transparent polymer. The polymer layer must be bonded on with a reflective adhesive. The part is then deformed under static conditions and illuminated with polarized light. Viewing the part with a special polarizing optic system will visually display lines of equal displacement. The strain field may also be recorded on film or video tape. As with holographic interferometry, the displacement must be transformed to strains and stresses. With 127 μm/cm sensitivity, the method produces accurate, quantitative results. Furthermore, the apparatus is moderately priced at approximately $15,000, see Figure 5.4.

Moire interferometry is accomplished by first bonding a clear polymer film, marked with closely spaced lines, to the specimen. A second sheet, similarly marked, is then placed over the first but without bonding or electrostatic coupling.

During incremental static loading, the bonded layer deforms while the second, unbonded layer, does not. Consequently, some lines will superimpose while others overlap (the Moire effect). Lines of equal displacement appear, which can be photographed or transformed to strains and

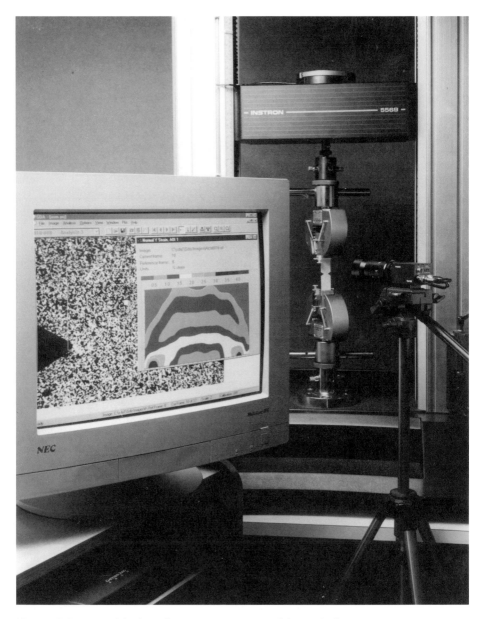

Figure 5.2 A speckle interferometer system. This optical system measures two-dimensional displacements of a component that has been painted with a speckle pattern.
(Courtesy of Instron Corp.)

Figure 5.3 A polymer cup with speckle paint. After hot water is poured in the cup, the interferometer calculates and superimposes two-dimensional displacement vectors showing the result of thermal expansion. A similar measurement could be made after application of mechanical load.
(Courtesy of Instron Corp.)

stresses. The technique is inexpensive (less than $1000 for equipment and materials), but limited in application because significant curvature cannot be accommodated.

Crackle lacquer coating entails painting the prototype with a brittle lacquer, then incrementally deforming the part. The lacquer cracks at a known strain, and the lacquer will develop easily observed regions of dense cracks when the failure strain of the lacquer is developed during incremental loading. The cracks are perpendicular to the direction of strain, thus providing information on the direction of maximum stress. Such lacquers are commercially available with failure strains from 0.0005 to 0.0008 cm/cm, as

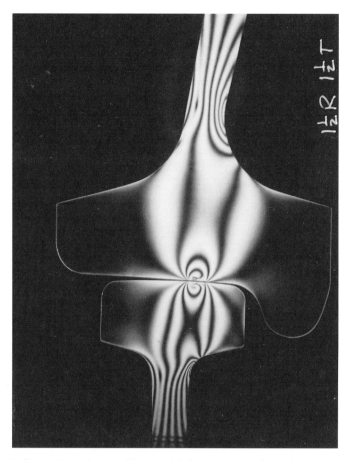

Figure 5.4 A photoelastic image. The model shows two surfaces in contact under load and the resulting two-dimensional displacement field.
(Courtesy of Instron Corp.)

determined by temperature and humidity conditions. This technique is inexpensive (under $1000 in equipment and materials) but produces limited information: it only indicates the location and approximate magnitude of stress concentrations. Furthermore, if lacquer cracks appear below routine service loads, an appropriate modification of the design may not be obvious because the relationship between the loading across the entire part and local strains is probably not linear. Crackle lacquer coating analysis, however, is useful for screening evaluations (such as proof loading where it is determined whether the prototype can sustain a typical service load) or determining suitable placement of strain gauges on parts with complex geometry.

Adhesively bonded resistance strain gauges provide an accurate, sensitive means of measuring strain at a point. They can be as small as 0.2 by 0.17 cm with a gauge length of 0.02 cm, and in "rosette" configuration can measure strain in two perpendicular directions. With proper attention to technique, one can obtain resolutions of 1 or 2 μm/cm, and record strain from static to 20 kHz transient conditions. Accuracy is typically 4% of the measured strain but can be better than 1% with special procedures. Standard gauges may be used at temperatures from -270 to $+290$ °C and up to 0.025 cm/cm strain. The apparatus is not expensive (about $3000), but some skill is required to adhesively bond the gauges to the part and make solder connections without causing damage to the gauge or the part. Strain gauges measure in-plane strains on surfaces even with curvature as high as 0.03 cm of radius.

Strain gauges cannot be feasibly placed over the entire surface of a component, so key points of high stress must be identified before instrumenting the prototype with gauges. In many cases, these critical areas are easily identified by judgment. Figure 5.5 shows several types of strain gauges.

Note that similar gauges may be used for sensing surface temperature at a point. This is most useful for prototyping localized heat transfer.

Single- or dual-spot speckle interferometry measures motion past a point without the need of physical contact. Using dual spots makes it possible to measure relative displacement by calculating the difference between the motions. Resolution is very high at 0.5 μm/cm, and motions of a wide range of rates may be observed. The prototype must have a diffuse reflecting surface; it may be necessary to apply an appropriate coating if the prototype does not already have suitable surface reflectivity. Laser speckle interferometers cost approximately $50,000 and have the disadvantage that only one point is monitored per interferometer. They are most useful in specialized prototyping analyses where high loading rates, high heating rates, or delicate parts are involved.

Optical extensometry measures relative displacement of two marks or targets. The resolution and rate of measurement depends on optics and the image algorithm, typically ranging from 5.1 μm to 0.51 μm with up to 30 data points per second. A basic type of optical extensometry is simple photography, but dedicated electronics and a computer will yield higher resolution and real time data handling (see Figure 5.6). Like speckle interferometry, the noncontacting nature of optical extensometry is particularly suited to tests with temperature fluctuations and delicate parts.

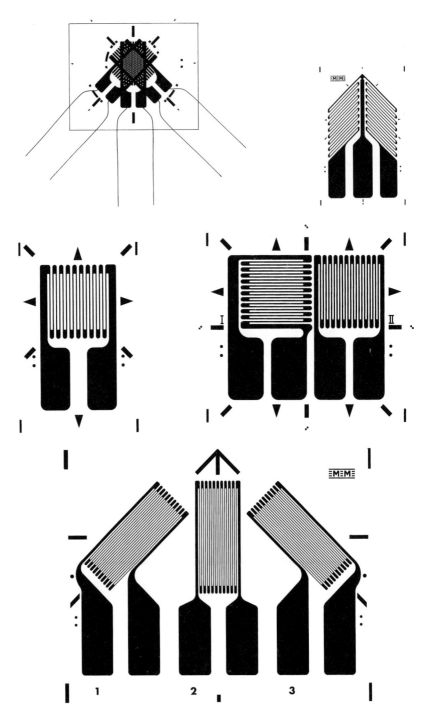

Figure 5.5 Several examples of the variety of adhesive-bonded strain gauge sizes and configurations that are available.
(Courtesy of Vishay Measurements Group)

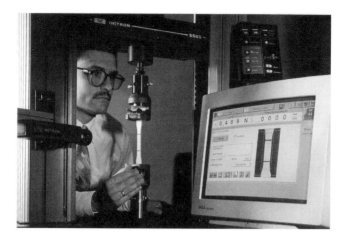

Figure 5.6 A noncontacting video extensometer. This device accurately measures relative displacement, along a single axis, of two optically contrasting marks on a test piece.
(Courtesy of Instrom Corp.)

5.4   Case Studies

The following case studies outline the process of prototyping applied to hypothetical applications.

5.4.1   Hand Drill Housing

Design requirements:

- impact resistance (in the event the drill is used in cold weather and dropped from a height),
- dimensional stability (to ensure correct alignment of motor and chuck shaft bearings over an expected range of end use temperatures),
- high stiffness at operating temperature to maintain alignment of bearings under stress at operating temperature, and
- minimum material and production costs.

The highest temperature that the housing will have to withstand is determined by the thermal conductivity of the housing because the maximum rate at which heat is generated by the motor is determined by the motor power rating. The conductivity of the housing is a complex function of the geometry, and finite element analysis models must be verified, justifying prototyping.

Based on cost, the applicable properties and processability, there appear to be two likely materials: nylon 6 and polypropylene, both 30% glass fiber reinforced.

Appropriate gating and venting are not easily predicted, so an aluminum prototype mold is required. The prototype molded parts are then inspected and subjected to service loads and temperatures in the laboratory.

Microscopic examination reveals successful molding without porosity.

Low- temperature impact strength is assessed by cooling the prototype drill assembly to $-40\ °C/°F$ and dropping the drill 2.4 m (8 ft) to a concrete floor, using a vertical tower with a guided carriage and a clamp to control the orientation of the drill. Side, end, and corner orientations are observed for each material and both perform satisfactorily.

Dimensional stability and stiffness at high temperature is measured by restricting cooling airflow partially and operating the drill at maximum torque for 5 min. Upon disassembly and inspection, it is evident that the polypropylene housing has undergone significant distortion while measurement of the nylon housing shows no bearing misalignment.

In conclusion, nylon is selected over polypropylene because it meets the functional requirements for this application and polypropylene does not. An illustration of this part is shown in Figure 5.7.

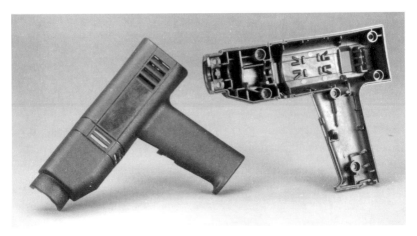

Figure 5.7 Hand drill housing
(Courtesy of DSM Engineering Plastics)

5.4.2 Electronic Instrument Housing

Design requirements:

- provide shielding against electromagnetic interference (EMI),
- protection of encased electronics from external sources of electrostatic discharge (ESD),
- sufficient stiffness and creep resistance to support 1.5 kg (3.3 lbs) of office items (e.g., manuals) stacked on top of the housing for several months at operating temperature, and
- minimum production and material cost.

Polycarbonate reinforced with stainless steel fibers was selected as a candidate material. Because the modulus of this material is upwards of 3445 MPa (500,000 psi), calculation shows that a wall thickness of 0.25 cm (0.10 in.) or more will withstand the low stress created by the manuals.

Several prototype housings are then produced, employing fan gating to minimize fiber attrition, from polycarbonate compounds containing stainless steel fiber concentrations of 5, 10, and 15%. Each prototype is assembled to house working electronics. Electromagnetic interference is generated with suitable tuned transmitters and electrostatic discharge applied with a static charge generator. EMI and ESD performance is acceptable at 10 and 15% fiber concentrations. The housings with 10% fiber reinforcement are loaded with 4.5 kg (10 lbs) of weight and left for 90 days with the unit turned on. Less than 0.12 cm (0.05 in.) of permanent deformation is measured when the housings are disassembled, which is expected to be acceptable to customers. This material is approved for production with 10% fiber concentration. An illustration of this part is shown in Figure 5.8.

Figure 5.8 Electronic instrument housing (Courtesy of DSM Engineering Plastics)

6 Specifications and Testing

Mitchell R. Jones

6.1 Specification Development

The development of sound, appropriate specifications is critical to manufacturing consistent, economical, high-quality parts made of short fiber reinforced plastic. It is essential that the design engineer know and understand applicable property measurement test methods and the use of statistics; it may also prove useful to have an understanding of microstructure and the variables that influence it. Specifications and testing programs serve to

- assess quality of raw materials,
- evaluate candidate materials and processing variables,
- provide design information,
- evolve prototype designs,
- reveal quality and consistency of manufactured parts, and
- define quality control/assurance programs.

Standardized test methods are most useful when writing a material specification for controlling the quality of incoming materials. Organizations that promulgate standardized test methods include:

- American Society for Testing and Materials (ASTM),
- United States Military Specifications (MIL),
- Federal Test Method Standards,
- Society of Automotive Engineers,
- Aeronautical Material Specifications,
- International Standards Organization (ISO),
- Deutscher Normenausschuss (DIN): German standards,
- British Standards Institution (BSI), and
- Japanese Industrial Standards (JIS).

Note that some standards allow several ways to measure a specific property. Further, be careful that all pertinent details of the specification have been addressed, including the processing and end-use application sections.

A common mistake in material specifications is to rely on composition values, particularly the percentage of reinforcement, for the purpose of controlling strength, modulus, toughness, and other properties. Although composition is often an important element of a quality control program, the critical factors are end-use properties, which ultimately determine whether a part performs or fails. Otherwise similar compositions produced by different vendors may well show significant physical property differences, making the attainment of a certain percentage of reinforcement less important than meeting a critical property minimum value, such as tensile strength or heat deflection temperature. Processing variations also have a significant influence on end-use properties; therefore, certain processing-related properties should also be specified, such as viscosity or melt index.

Finished part specifications are not usually based on standardized test methods and are typically developed specifically around part performance requirements. Note the following considerations:

- Dimensions are often a critical element, but costs rise rapidly with increasingly small tolerances.
- Other performance tests are often developed during the prototyping and testing process, particularly component impact tests.
- Part appearance is often specified, especially if color matching is involved; the requirements should be quantified, rather than based on subjective judgement, so that the process will deliver acceptable quality parts, on time, at a competitive cost.
- Statistical process quality control (SPQC) should be incorporated in the development of specifications, to help detect and prevent quality problems before they occur. The measurement and analysis of trends indicates the importance of consistency in the course of manufacturing.

In summary, specification development consists of the following elements:

- choosing critical, measurable properties relevant to the application and the customer's definition of quality;
- correlating/validating those properties with end-use performance;
- establishing the test method and the desired range of values obtained through the test (allowing for the precision and reproducibility of the test);
- generating feedback from the comparison of test results with end-use performance to ensure that the specifications are realistic and cost effective.

Table 6.1 lists the principal properties of short fiber reinforced plastics that may be considered for specifications and Table 6.2 lists the ASTM test methods that may be applied to measuring these properties. It seems likely that eventually ASTM test methods will be harmonized with ISO test

Table 6.1 Properties of Short Fiber Reinforced Plastics

I. Physical properties
  A. Mechanical properties
    1. Mode of deformation
       a. Tension
       b. Compression
       c. Flexure
       d. Torsion
       e. Shear

    2. Material properties
       a. Modulus of elasticity (elastic stiffness)
       b. Yield stress (stress that induces nonelastic behavior or significant permanent deformation)
       c. Ultimate stress
       d. Ultimate strain
       e. Toughness (total energy absorption)
       f. Notch sensitivity (resistance to crack initiation)
       g. Fracture toughness (resistance to crack growth under monotonic loading)
       h. Cyclic fatigue life

    3. Variables
       a. Rate of loading (long-term creep → impact)
       b. Environment
          i. Temperature and thermal history
          ii. Humidity
          iii. Presence of corrosives or solvents
       c. Repetition of loads (monotonic loading → cyclic loading)

    4. Other mechanical properties
       a. Wear resistance
       b. Surface indentation hardness
       c. Dynamic modulus (vibration energy absorption and visoelastic properties)

  B. Thermal properties
    1. Coefficient of thermal expansion
    2. Thermal shrinkage
    3. Thermal conductivity
    4. Specific heat ("thermal mass")
    5. Brittleness temperature
    6. Heat distortion temperature (deflection under load)
    7. Glass transition temperature

Table 6.1 *(Continued)*

    C. Electrical properties
        1. Electrical resistance/impedance (surface and volume)
        2. Dielectric strength (breakdown voltage)
        3. Dielectric constant and power loss factor
        4. Arc resistance

    D. Other physical properties
        1. Density
        2. Void content (porosity)
        3. Machinability
        4. Magnetic properties
        5. Flammability, ignition resistance, smoke generation, and flame spread

II. Chemical properties
    A. Resistance to chemical agents
        1. Acids and bases
        2. Solvents and fuels
            a. Change in physical or optical properties
            b. Change in chemical composition
        3. Moisture
        4. Biological
        5. Corrosive or oxidizing gases

    B. Resistance to radiation—sunlight and other sources

    C. Toxicity
        1. Outgassing (loss of volatile gases)
        2. Presence of toxic chemicals on the surface

    D. Environmental
        1. Emissions to water or atmosphere
        2. Worker exposure to foregoing during fabrication

III. Appearance and optical properties
    A. Scratch and abrasion resistance
    B. Mar resistance
    C. Color
    D. Surface smoothness
    E. Transmittance, reflectance, absorption, and index of refraction
    F. Distortion or warparge
    G. Crazing resistance

Table 6.2 List of ASTM Standards Applicable to Testing Short Fiber Reinforced Plastic

Test Method	ASTM Standard Number
Abrasion wear	D 1242
Bearing strength	D 953 Method A
Bending modulus (cantilever beam)	D 747
Blocking	D 1893
Bond or cohesive strength (thickness direction tensile strength)	D 952
Bonding strength (strength in thickness direction)	D 229
Brittleness temperature of plastics by impact	D 746, D 1790
Calibration of durometers (surface hardness)	D 1706
Chemical resistance of thermoset resins	C 581
Chip impact strength	D 4508
Classification and specification system	D 4000
Classifying visual defects	D 2562, D 2563
Coefficient of cubical thermal expansion	D 864
Coefficient of friction	D 1894, D 3028
Coefficient of thermal expansion	D 696, E 831, E 289
Color	D 1729
Colorfastness to light	D 795
Compressive properties	D 695
Compressive properties	D 1621
Conditioning methods for testing	D 618
Corrosion	D 4350
Creep properties	D 2990
Definitions of terms relating to plastics	D 883, D 4092, E 6, E 41, E 375
Deflection temperature under load	D 648
Deformation under load	D 621
Density	D 792, D 1622
Density of smoke from burning	D 2843
Determining mold temperature	D 957, D 958
Dielectric breakdown voltage and strength	D 149
Dissipation factor and dielectric constant	D 150, D 669
Dynamic mechanical properties (loss modulus)	D 4065
Electric arc resistance	D 495
Electrical resistance	D 257
Extinguishing characteristics	D 3801
Fatigue (flexural)	D 671
Flammability	D 568, D 635
Flexural properties	D 790
Flow temperature for molding	D 569
Gloss (appearance)	D 523
Heat aging	D 3045
Heat distortion temperature (tensile)	D 1637
Ignition	D 1929, D 3713, D 3894
Ignition loss	D 2584
Impact resistance (falling weight)	D 3029
Indentation hardness (Barcol)	D 2583

Table 6.2 *(Continued)*

Test Method	ASTM Standard Number
Indentation hardness (Rockwell)	D 785
Index of refraction	D 542
Izod impact strength	D 256
Instrumental impact test	D 3763
Light (carbon or xenon arc; fluorescent UV; concentrated natural sunlight) and water resistance	D 1499, G 23, D 2565, G 26, D 4459, G 53, D 4329, D 4364
Light diffusion	E 166, E 167
Linear dimensional changes under accelerated service conditions	D 1042
Moisture content	D 4001
Molecular weight	D 3592, D 3593, D 3750, D 4001
Outdoor weathering	D 1435
Oxygen index (minimum for burning)	D 2863
Particle size (sieve analysis)	D 1921
Permanent effect of heat	D 794
Resistance to bacteria	G 22
Resistance to chemical reagents	D 543
Resistance to fungi	G 21
Resistance to sulfide staining	D 1712
Rheological properties by capillary	D 3835
Rheological properties by dynamic mechanical analysis	D 4440
Shear strength under punching operations	D 732, D 3763, D 617
Shear strength, in-plane	D 3846
Shrinkage after molding	D 955
Smoke particulate determination	D 4100
Stain resistance	D 2299
Stiffness in torsion at various temperatures	D 1043
Stress relaxation	D 2991
Tensile impact energy	D 1822
Tensile properties	D 638, D 1708, D 1623, D 2289
Test specimen mold design	D 647
Thermal conductivity	D 4351
Transition temperatures (thermal analysis)	D 3418
Transmittance of reinforced panels	D 1484
Void content (porosity)	D 2734
Volatile determination	D 4526
Volatile loss	D 1203
Water absorption	D 2842, D 570
Water vapor transmission	E 96
Weight and shape changes under accelerated service testing	D 756
Yellowness index	D 1925

methods, which are more definitive than ASTM methods. However, other than for some automotive applications, ISO standards are not in widespread use at present in the United States.

6.2 Total Quality Management

With benefits that penetrate every level of business and society, Total Quality Management (TQM) has prevailed as an essential part of any successful company's philosophy. TQM entails a commitment to excellence from every employee, at every level, without compromise; the strategy is to optimize the process of supplying a good or service to prevent problems from ever occurring.

It is implemented through several principles:

- Every individual is both a supplier and a customer. Goods and services are provided and accepted only when the customer is satisfied.
- All phases of the product are examined. From its inception, the product must be robust, tolerating common variances in manufacturing and use.
- The ideal to strive for is zero defects, 100% yield, and zero product variation.

The result is giving customers what they want, with benefits of optimal profit and employee satisfaction for all parties. Companies that implement TQM earlier will secure more favorable market position.

Once implemented, TQM can facilitate significant cost reductions. On both the macro scale (between companies) and micro scale (between individuals), warranty, scrap, repair, analysis, inspection, and testing are reduced or eliminated. For example, integrated suppliers do not force the customer to test incoming products for quality—it is assured.

Implementation must occur in several areas: culture, quality control methods, and method of design and process experimentation.

6.2.1 Culture

First, management must accept responsibility for success or failure of the complete system and must provide the leadership, working conditions, and job security to support each individual's pursuit of increasing customer satisfaction. Next, every worker must accept the uncompromising objective of customer satisfaction, meaning fulfillment of the customers' needs and satisfaction in using the product. *Each worker must also strive for continuous improvement.*

Another difficult cultural step is adoption of statistical thinking. Each worker must understand and apply measurement and effects of variation. Variation effects product design, production, and the customer. An example of this thinking is the "quality loss function" which relates the increasing financial loss to a customer resulting from deviation of the product from the target properties.

6.2.2   Statistical Control

The simplicity and effectiveness of Precontrol make it the preferred method for statistical quality and process control.

Based on statistical theory, it does not suffer shortcomings of traditional statistical process control techniques; Precontrol rarely stops a process that should be allowed to continue nor does it permit poor quality.

Precontrol is implemented in four easy steps:

1. Green, yellow, and red zones are defined on a control chart by calculating quartiles in relation to a pair of specification limits or a design target and a lower specification limit. The green zone is the two center quartiles; the yellow zones are the outer quartiles; and the red zones are the regions outside of the specification limits (outside of the four quartiles). These conditions are shown in Fig. 6.1.
2. Determine whether the process maintains acceptable variability. Do this by taking at least five initial units. If all fall in their green zone, production commences. If any fall in the yellow or red zones, the variation is not acceptable and corrective action is taken. Engineering judgment or Design of Experiments (DOE, explained below) are employed until all units are in the green zone, whereafter production commences. Again, note the illustration of "go-no go" in Fig. 6.1.
3. Two consecutive units are periodically taken from production. The following rules are observed:
   - Production continues if
     - both units are in the green zone or
     - one unit is in the green zone and one in either yellow zone.
   - Stop production if
     - both units are in one or both yellow zones
     - one or both units are in either red zone

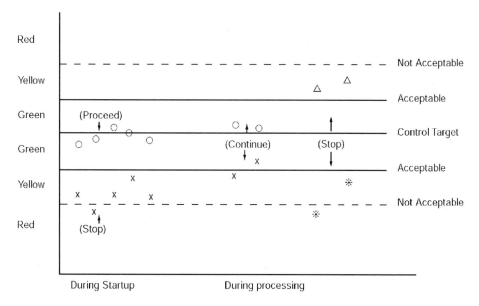

Figure 6.1 Precontrol chart examples

Figure 6.1 shows these situations. If production is stopped, again engineering judgment or DOE are applied to reduce or eliminate the cause of excessive variation. Step two is reapplied next.
4. The period of time between sampling two consecutive units is optimized by:
   - engineering judgment initially and
   - thereafter dividing the number of units produced between stoppages by six to determine the sampling period.

The last step—determining sampling period—is a special feature of Precontrol; it forces improvement of a poorly controlled process and rewards a well-controlled, low-variance process.

Finally, it is noteworthy that a production worker can easily chart and understand the simple sampling and categorizing used in Precontrol. Even a pocket calculator is unnecessary.

6.2.3　Reduce Variation

As mentioned earlier, a goal of TQM is reduced product variation. This is accomplished by use of sound engineering judgment and Design of Experiments (DOE) during product and process design and when resolving recurrent manufacturing problems.

Variation is ideally reduced (or eliminated) when the product and manufacturing process are designed. If the design is robust, common process variances and likely conditions in product use will not cause reject parts or failures.

Whether used for design or to correct a faulty process, judgment and DOE pursue the following objectives:

1. Identify the most influential variables, whether they are product or process parameters, purchased goods, or measurement variables.
2. Rank the variables from most influential to least.
3. Reduce or eliminate the variation caused by the most influential variables through close tolerances, conservative design, more careful process control, or whatever means are effective. Pay attention to interaction effects and pursue the alternatives with greatest economic benefit (including consideration for cost to the customer of a failure in service and other "hidden" costs of permitting variation). Typically, four or fewer variables are highly influential.
4. Reduce costs by using liberal tolerance, avoiding conservative design, using less precise process control, etc., for the least influential variables.

The diagnostic DOE procedures follows those described by Taguchi [1]. This stepwise, relatively uncomplicated series of experiments reliably indentifies the influential variables and their interdependence with a minimum time investment.

The implementation of TQM should not be considered a substitute for ISO 9000 certification, or vice versa. In fact, ISO 9000 certification should be an integral part of TQM. ISO 9000 requires that a company establish, document, and maintain a quality program, according to specified standards, and that the company be audited regularly to ensure compliance with its own documented program in accordance with these standards. Therefore, successful ISO 9000 certification will create most of the foundation for TQM but does not close the loop by itself between customer satisfaction and company responsiveness. This is the critical gap that must be filled by the commitment of management to TQM.

Total Quality Management is the only business strategy that still succeeds in the world industrial market. It generates maximum profits and employee satisfaction in a reasonable amount of time. With backing from executive management, simple philosophies, cultural changes, and analytical tools are the only requirements.

References/Bibliography

1. Bendell, T. (ed.) "Taguchi Methods," Proceedings of the First European Conference on Taguchi Methods, London, UK, 13–14 July 1988
2. Agarwal, B.D., Broutman, L.J. *Analysis and Performance of Fiber Composites* (1980) John Wiley, New York
3. Athalye, A.S. *Plastic Materials Handbook* (1980) Multi-Tech Publishing, Bombay
4. Berins, M.L. (ed.) *SPI Plastics Engineering Handbook of the Society of the Plastics Industry, Inc.* (1991) Chapman and Hall, New York
5. Brown, R.P. (ed.) *Handbook of Plastics Test Methods* (1986) Longman Scientific and Technical, Harlow, Essex, England
6. Clegg, D.W., Collyer, A.A. *Mechanical Properties of Reinforced Thermoplastics* (1986) Elsevier Applied Science, New York
7. Datoo, M.H. *Mechanics of Fibrous Composites* (1991) Elsevier Applied Science, New York
8. Folkes, M.J. *Short Fibre Reinforced Thermoplastics* (1982) Research Studies Press, Chichester, Sussex, England
9. Harper, C.A. (ed.) *Handbook of Plastics, Elastomers, and Composites* (1992) McGraw-Hill, New York
10. Holmes, M. *GRP in Structural Engineering* (1983) Applied Science, London
11. Leggatt, A. *GRP and Buildings: A Design Guide for Architects and Engineers* (1984) Butterworths, London
12. Lincoln, B., Gomes, K.J., Braden, J.F. *Mechanical Fastening of Plastics: an Engineering Handbook* (1984) Marcel Dekker, New York
13. Kobayashi, A.S. (ed.) *Handbook on Experimental Mechanics* (1987) Prentice-Hall, Englewood Cliffs, NJ
14. Mallick, P.K., Newman, S. (eds.) *Composite Materials Technology: Processes and Properties* (1990) Hanser, Munich
15. Matthews, F.L. (ed.) *Joining Fibre-Reinforced Plastics,* (1987) Elsevier Applied Science, New York
16. Milewski, J.V., H.S. Katz, (eds.) *Handbook of Reinforcements for Plastics* (1987) Van Nostrand Reinhold, New York
17. Miller, E. (ed.) *Plastics Products Design Handbook* (1983) Marcel Dekker, New York
18. Mohr, J.G. (ed.) *SPI Handbook of Technology and Engineering of Reinforced Plastics/Composites* (1981) R. E. Krieger, Huntington, NY
19. Nielsen, L.E., Landel, R.F. *Mechanical Properties of Polymers and Composites* (1994) Marcel Dekker, New York
20. Rosato, D.V. *Plastics Processing Data Handbook* (1990) Van Nostrand Reinhold, New York
21. Rubin, I.I. (ed.) *Handbook of Plastic Materials and Technology* (1990) Wiley, New York
22. Saechtling, H. *International Plastics Handbook: for the Technologist, Engineer, and User* (1987) Hanser, Munich
23. Shenoi, R.A., Wellicome, J.F. (eds.) *Composite Materials in Maritime Structures* (1993) Cambridge University Press, New York
24. Shook, G.D. *Reinforced Plastics for Commercial Composites* (1986) ASM, Metals Park, OH

25. Summerscales, J. (ed.) *Non-destructive Testing of Fibre-reinforced Plastics Composites* (1987) Elsevier Applied Science, New York
26. Thayer, A.M. *Chemical Companies Extend Total Quality Management Boundries* (1995) *Chem. Eng News,* 27 Feb. 1995
27. Ward, I.M., Hadley, D.W. *An Introduction to the Mechanical Properties of Solid Polymers* (1993) John Wiley, New York
28. Weatherhead, R.G. *FRP Technology: Fibre Reinforced Resin Systems* (1980) Applied Science Publishers, London
29. Wool, R.P. *Polymer Interfaces: Structure and Strength* (1995) Hanser, Munich
30. Staff of *Modern Plastics, Plastics Handbook* (1994) McGraw-Hill, New York
31. *Composites of Engineered Materials Handbook,* Vol. 1, (1995) American Society for Materials

Additional Sources of Information

American Society of Mechanical Engineers
American Society for Materials
American Society for Testing and Materials
Plastics and Rubber Institute
Society of Plastics Engineers
Society for Experimental Mechanics
Society of Automotive Engineers
Society of the Plastics Industry

7 Injection Molding

Donald V. Rosato

7.1 Introduction

Although many fabricating processes are employed to produce unreinforced plastic products, at least 50% by weight of short fiber reinforced plastics go through injection molding machines. For small quantities of material, however, other processes can compete with injection molding (IM), among them being compression molding, transfer molding, casting, and thermoforming [1 to 13].

IM has the advantage that molded parts can be manufactured economically in virtually unlimited quantities with little or practically no post-molding finishing operations. It is principally a mass-production method, and because of the relatively high capital investment in machines, molds, and auxiliary equipment, the best return on this investment is realized when it is so used [2]. The surfaces of injection moldings are as smooth and bright or grained and engraved as the surfaces of the mold cavity. IM is characterized not only by its rapid growth in output, but also by its reliability and versatility, processing new types of materials to close tolerances for all types of applications. There is a practical and easy approach to injection molding with plastics; it is essentially no more difficult than fabricating with other materials such as metal, glass, or wood. IM plastic parts have been produced for more than a century (IM of plastics is believed to have begun in 1872). They have been used successfully and have provided exceptional cost advantages, compared to parts fabricated from conventional materials. Ironically, some people think that IM plastics are new because the industry has an endless capability of producing new compounds to meet new performance requirements or they are not familiar with the IM process [1 to 6].

7.2 Basic Process Description

IM is a simplified name for a quite complicated process (complicated to one unfamiliar with its basic operation or unfamiliar with how to design parts and molds properly), that is controllable within specific limits. The concept of "within limits" is similar to other manufacturing processes [2].

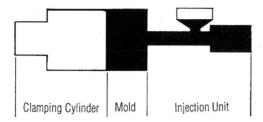

Figure 7.1 Basic elements of the injection molding process

The IM process involves introducing the material into a heating chamber where the compound melts (the term in common usage and the one used hereinafter for convenience, although, technically, it *softens* if it is amorphous or *melts* if it is semicrystalline), then injecting it under pressure into a matched-metal cavity. The part then solidifies into its intended shape. With thermoplastic compounds, solidification occurs by *cooling* the melt (molten plastic), but with thermoset compounds, solidification occurs by *heating* the melt in the mold cavity (to achieve polymerization and cross-linking) [1]. The next step involves ejecting the part from the mold. As shown in Fig. 7.1, three basic mechanical units are combined to perform injection molding: the melting/injection unit, the mold and a mold clamping device.

The mold may consist of a single cavity with a sprue that channels the flow of the melt from the heating/injection chamber to the mold cavity. Molds may also have multiple cavities, either similar or dissimilar, each connected to flow channels (runners) that direct the flow of the melt from the sprue to the cavities (Figs. 7.2 through 7.6).

7.3   Machine Operation

Under practical operating conditions, the process of learning IM takes place in three stages [2]:

1. Stage 1 concerns the actual operation of the machine.
2. Stage 2 involves setting machine conditions to prescribed values to melt and control the temperature of the material as it moves from the feed hopper through the cylinder, transfers into a properly designed mold, and is transformed into acceptable parts.

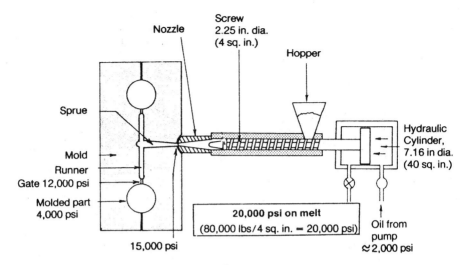

Figure 7.2 Example of pressure loading on the melted plastic going from the plasticator (injection unit) to the mold (right to left)

3. The final stage is devoted to problem solving and fine tuning the process, leading to optimum productivity and that quality of parts that meets performance requirements at the lowest cost.

Plastic materials are transferred from their containers, for example, bags or boxes, to a feed hopper from which they move by gravity flow to the throat of the cylinder (or "barrel") of the reciprocating screw IM machine; many resins must be dried before they are placed in the feed hopper (see Section 7.12). The granules are then typically picked up by the flights of a rotating screw and conveyed into the heating section of the cylinder. For heavy, high-viscosity plastics with high reinforcement content of long fibers, such as bulk molding compounds (BMC), a ram is often used, or a combination of ram followed with a screw motion. Figures 7.5 and 7.6 include some of the basic melting systems. The conveying action of the screw/ram compresses the material as it moves into the heating section. There the material softens to such a degree that it becomes relatively fluid and is then conveyed to the front section of the cylinder, known as the measuring section. This action provides reasonably accurate "shot size" control of the amount of melt going into the mold. After the desired shot size is accumulated, the screw stops rotating and moves forward at a controlled rate, thus acting as a ram or plunger.

Raising the temperature of the plastic to the point where it will flow

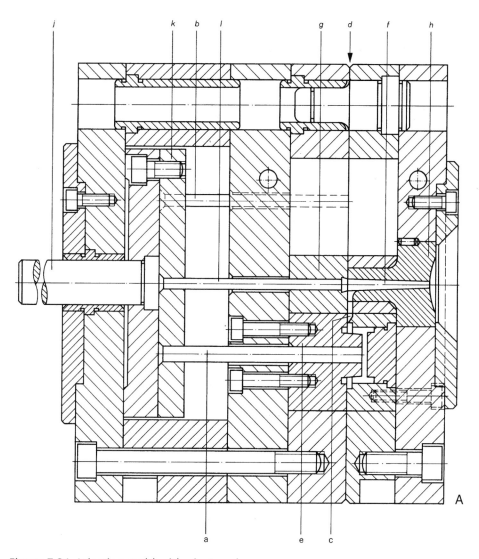

Figure 7.3A Injection mold with ejector pins

a	ejector pin	f	sprue
b	return pin	g	cavity mold plate
c	runner	h	sprue bush
d	parting line	j	stop bolt
e	cavity and rear	k	ejector plate
	cavity mold plate	l	ejector pin

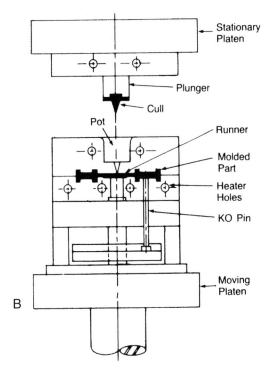

Stationary
Platen

Plunger

Cull

Pot

Runner

Molded
Part

Heater
Holes

KO Pin

Moving
Platen

B

Figure 7.3B Pot type transfer mold (PTTM)

under pressure is critical to ensure meeting such performance requirements as part dimensions, weight, toughness, surface finish, etc. This is normally done by simultaneously heating and masticating the plastic until it forms a melt possessing uniform temperature and viscosity. Where possible, modern reciprocating screw IM machines are preferred over older machines having only a ram or plunger because of the greater uniformity of the melt produced. This overall phase of IM is called plasticizing or plastication.

Next, the melt is forced into the mold cavity under controlled pressure (dictated by the viscosity of the melt as it flows and the geometry of the mold). As an example, the material shown in Figure 7.2 requires 27.6 MPa (4000 psi) pressure in the cavity. Compounds in commercial use today may require from 13.7 MPa (2000 psi) to 103.4 MPa (15,000 psi) or even up to 172.4 MPa (25,000 psi) cavity pressure. To meet the required cavity pressure, the plasticizing capacity of the machine, its clamping force and the dimensions of the mold cavities and runner system must be interrelated and designed to yield the correct cavity pressure. This is not as difficult as it may

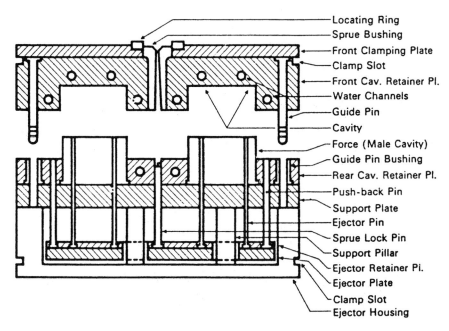

Figure 7.4 Exploded view of mold. This mold is in a vertical position, such as it would appear sitting on a table or being used in a machine with vertical clamps. However, most machines use horizontal clamps. Thus this mold would be rotated 90° and would be positioned similar to those in Figs. 7.1, 7.2, and 7.3.

appear because it has been done successfully on a largely empirical basis for over a century. Here is where theory and practice meet, requiring an understanding of plastics melt flow characteristics as well as how to set up and operate the different controls on an IM machine.

The molten plastic from the injection unit is transferred through the runner system to the cavities, where it is shaped into the desired form and held under the required clamp force until it has solidified. What makes this apparently simple operation complex is the limitations of our knowledge of the process necessary to fill all of the cavities evenly while the material is still cooling and solidifying.

With thermoplastics compounds, the mold is kept relatively cool, below the melt temperature. The time the melt takes to solidify ("setup") is important; with shorter setup times, more parts can be molded in a given period of time and unit production costs are reduced. An illustration of the components of a molding cycle for a thermoplastic is shown in Fig. 7.7. Also shown in the same figure are the components of a thermoset molding cycle [8].

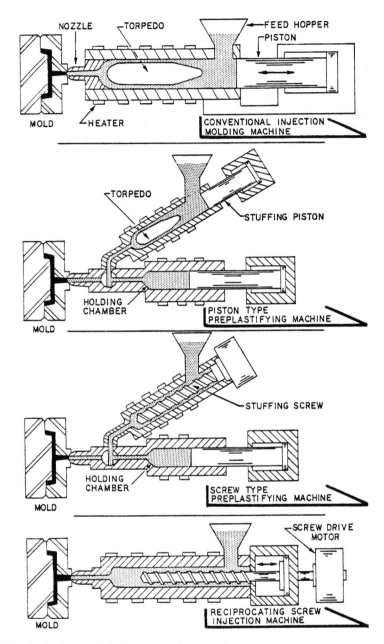

Figure 7.5 Basic different injection molding systems

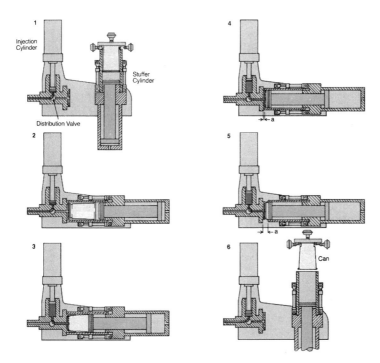

Figure 7.6 Example of injection molding machine for BMC using a can that contains the BMC and the stuffer cylinder (ram or plunger concept). Can is loaded into the stuffer cylinder—note that the distributor valve is still switched on injection into the cavity from the previous shot. Steps 2 through 5 show the can being compressed. in step 3, the distributor valve has switched over for material feed into the injection cylinder. The space (letter "a") opens up after the last shot, as shown in step 5, to allow for the can to drop oput. In step 6, the stuffer has returned to its original position. (Fjellman Press AB, Mariestad, Sweden, Ref. 1).

Note that the "cooling cycle" for a thermoplastic (shown in Fig. 7.7 as 25 s in the second line from the top) is called a "heating cycle" for a thermoset. In both cases, the term "cure time" is commonly used, even though it is not technically correct in the case of thermoplastics. The reason for this blurring of distinction is, that in the beginning of this century, practically all plastics were thermosets and the term (indicating solidification) carried over when the use of thermoplastics became widespread [1]. At the end of the cure time, the mold is opened, followed by ejection of the part using knockout pins (Figs. 7.3 and 7.4), a stripper plate, unscrewing core, etc. All of these operations and their controls on the machine can be easily understood, with the use of training manuals and schools (a few days to a week). This will

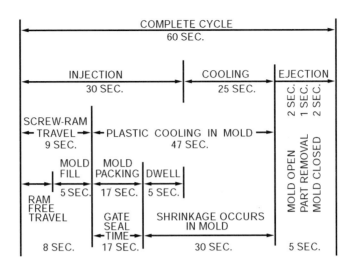

Figure 7.7 Breakdown of injection molding cycle for thermoplastic compounds. With thermoset compounds the main difference concerns the terminology of 25 s cooling time (second line down) where it should be identified as 25 s heating time.

enable a novice to get started, with fewer bad results and less time wasted than relying exclusively on cut-and-try experimentation. Too often, training time is not provided.

Usually, an IM machine is categorized by clamp force expressed in tons, and shot size—the maximum weight of plastic (normally, general purpose unreinforced polystyrene)—that can be plasticized and injected in one cycle. Obviously, shot size is an imprecise number, because this will vary with the density of the material being molded; therefore, shot volume would be a more useful parameter. Other machine parameters include injection rate, screw recovery rate (how long does it take to plasticize material for the next shot—important because it is not likely to be the same as the cure time), back pressure, screw speed, length-to-diameter ratio of the screw, screw design/shape, materials of construction for the screw and barrel, clamp stroke, daylight opening (the maximum dimensions of a mold that can be fitted into the machine), hydraulic line pressure, and delivery rate.

The operation of the machine depends on the characteristics of the material being processed. Each plastic compound differs in its ability to flow under heat and pressure, as well as its ability to flow within the confines of the runners and cavities of the mold. The recommendations of the material supplier for machine settings should be used as the starting point; trial-and-error is used to modify the settings until the user is satisfied with the qualities of parts produced and the cycle times required. Although experience will help one to cut the trial-and-error process short, training will help one to reach optimum results faster and without the waste of material incurred by mistakes.

Some materials and molds may require that both the speed of injection and injection pressure vary during the filling process. A heat-sensitive plastic may be degraded if too rapid a fill rate is used. Forcing the melt through small orifices, such as at the tip of the nozzle or small gates, at too high a velocity may also increase the internal shear (which will raise the melt temperature, particularly at the center of the channel) enough to cause overheating and "burning" (discoloration and degradation of the material); possible excessive attrition of the fiber reinforcement lengths is likely to result. For this reason, nozzles should have an inside diameter of 1.11 to 1.27 cm (0.4375 to 0.5000 in.), with an exit orifice of 0.635 cm (0.25 in.) and a land with a reverse taper of 2 to 3°.

However, thin-wall parts require a fast fill rate to prevent solidification of the material before the cavity has been properly filled and packed; this is particularly true of short fiber reinforced plastics to minimize orientation and enhance weld line integrity. Some molded parts have both thick and thin sections, plus interrupted flow patterns, resulting from the presence of cored holes. Such demanding requirements necessitate considerable versatility in the capability of the injection unit. Programming different injection speeds and pressures during the forward travel of the screw or plunger/ram can greatly aid in filling cavities properly. Programmable or multistage injection is standard on most IM machines built over the past 25 years; however, there are many machines older than this still in operation!

The clamp of a machine must have sufficient locking force to resist the tendency of the molten plastic moving at high pressures to force apart the mold halves. Again, this problem is aggravated when molding short fiber reinforced plastics because their reduced flow properties (compared to unreinforced materials) mean that higher injection pressures must be employed to push them into the mold [2, 4, 8]. If the mating surfaces of the mold are forced apart, molten plastic will squeeze out through the aperture, known as "flash." Thermoplastic compounds will not normally flash unless the mold opens more than a few thousandths of a centimeter; see Fig. 7.8 at location "A," where the cavity is vented. The mold cavity contains air, of course, that must be removed or displaced as the melt moves into the cavity. At high

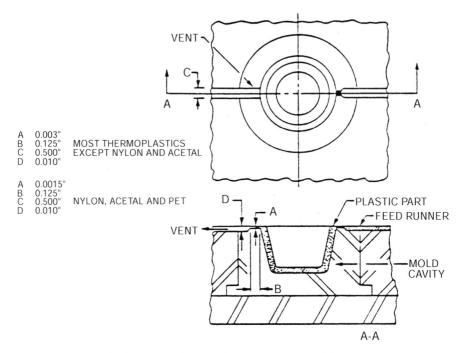

Figure 7.8 Method of venting thermoplastic injection molds

injection speeds, insufficient venting of the air present may result in considerable compression of the air, with consequent slow filling, premature buildup of back pressure and, in some cases, "burning" of heat-sensitive material at the melt-flow front due to the heat of the compressed air.

With thermosets, the material viscosity is much lower than is the case with thermoplastics, and some flashing is to be expected even with negligible forcing apart of the mold mating surfaces; it can only be minimized. Overflow vents (Fig. 7.9) and other techniques are used [8].

Usually, the amount of clamp force required depends on the projected area of the molded part and the runner system, as measured on the plane surface through the parting line of the mold (where the mating mold "halves" meet). This area multiplied by the pressure required to inject the melt and a typical safety factor of, say, two, equals the required clamp pressure expressed as tonnage. The clamping mechanism is generally provided by a locking toggle action, a hydraulic cylinder, or a combination of the two.

In the course of molding, some generation of fumes from the molten polymer is unavoidable. A good ventilating system, connected to a fume

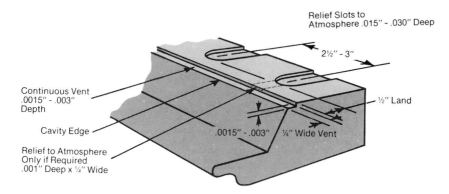

Figure 7.9 Example of applying a venting system at the mold parting line; similar to what has been used in various molds for thermosets (particularly phenolics) and structural foam injection molds (reinforced or unreinforced)

scrubber, is important for the safety of the plant operators, as well as the area environment. Although none of the polymers discussed in this book are known to evolve toxic fumes under normal conditions, care must taken to avoid overheating resins to the point of decomposition. A Material Safety Data Sheet (MSDS) must be provided by law to any new buyer of a product. The processor should read the MSDS carefully to learn if there are any special precautions needed if the polymer decomposes.

7.4 Types of Molds

There are basically two different types of molds. One type requires that the sprue and runners solidify as the part does, so that the sprue and runners are removed when the part is ejected. Figure 7.10 shows a "two-plate" type mold for thermoplastics, containing water lines that cool both the part and the runner; hence the name "cold runner" mold. When the mold is kept hot (electrical heaters, hot oil, etc.) around the sprue and runner, and water cooling lines used around the cavity, only the plastic in the cavity solidifies. This type is called a "hot runner" mold; it is more difficult to operate but much more efficient than a cold runner mold [2, 4, 7].

With thermosets, the terminology is reversed. It is necessary to maintain the mold at a higher temperature than the melt so that the plastic will polymerize and solidify. The runners also cure and solidify in a typical two

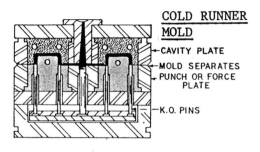

COLD RUNNER MOLD

- CAVITY PLATE
- MOLD SEPARATES
- PUNCH OR FORCE PLATE
- K.O. PINS

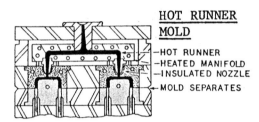

HOT RUNNER MOLD

- HOT RUNNER
- HEATED MANIFOLD
- INSULATED NOZZLE
- MOLD SEPARATES

Figure 7.10 These diagrams represent the cold runner mold and hot runner mold when using thermoplastic molding compounds. With thermoset molding compounds basically the top diagram is a hot runner mold and the bottom is a cold runner mold (explanation is given in the chapter text).

plate mold, so that we now have a "hot runner" mold for thermosets operating in the same fashion that a "cold runner" mold for thermoplastics would.

The opposite of a "hot runner" for thermoplastics would be a "cold runner" for thermosets. The runner area is kept relatively cool (but still hot enough to ensure flow) while the area around the cavities is kept at a higher temperature than the melt. Thus only the parts solidify. These types of molds have been used successfully for many years, although, again, they require more sophistication to operate than one that produces solid runners [2].

In addition to these two basic types, there are many others that are derived from them. These include, among others, three plate molds, insulated runner molds, hot manifold molds, and stacked molds. Note that in a single cavity mold no runner is present, only a sprue (Fig. 7.4). Obviously, productivity and hence economics are enhanced if one can eliminate making solid sprues and runners that must be recycled or scrapped, and that add to cycle time and shot weight as well. With thermoplastics, the solid sprues and runners can be granulated (and become known as "regrind") and remolded, usually as a blend with "virgin" (unprocessed) material. Most material suppliers recommend not using more than 20 to 25% of such reworked material, because each time it is remelted, some reduction of properties (e.g., strength, toughness, appearance) is likely to take place. Short fiber reinforced thermoplastics are uniquely affected adversely by recycling because both granulation of the scrap and replasticizing cause attrition of the fiber length, with attendant diminishment of strength and toughness, etc.; thus precautions associated with the use of regrind must be rigorously followed [1, 2, 10, 12]. Disposal of any unusable scrap or regrind is discussed in Chapter 4, Section 4.10. Nevertheless, with proper mold design and careful processing, it should be possible to achieve virtually 100% material utilization.

For extended production runs, the use of harder steel (up to 350 Brinnel hardness) or chrome plated cavities are recommended for increased tool life when molding glass fiber reinforced materials. Low sulfur/low phosphorus-containing steels are preferred when molding compounds containing high loadings of molybdenum disulfide [2, 3]. As noted in Chapter 5, aluminum and aluminum-filled epoxy molds are satisfactory for prototype work.

The sprue bushing should be as short as possible and have a taper of at least 3°. The orifice should be slightly larger than the nozzle; for example, when using a 0.635 cm (0.25 in.) nozzle, the sprue bushing should have an orifice of 0.559 cm (0.22 in.). Of course, the radius of the nozzle and sprue bushing must match; otherwise, the sprue may not eject properly from the mold.

Runners for short fiber reinforced plastics are preferably full-round or trapezoidal in cross-section, to ensure full flow.

7.5   Types of Screw Injection Units

Whenever possible, a screw plasticizing system should be used. The typical reciprocating screw will melt and mix the plastic (which has been heated by contact with the hot walls of the barrel) through compression and shear, and as the screw moves to the rear of the cylinder, the melt accumulates in front of the screw. The screw stops rotating when the desired shot volume is attained and then the screw, acting as a ram, pushes the melt into the mold (Fig. 7.11).

The main components of an injection unit are the screw located within a cylindrical barrel, a motor to rotate the screw, a device to move the screw as a ram (such as a hydraulic cylinder at the back of the screw), and controls to operate (rotational speed, shot size, temperature, pressure, time, etc.).

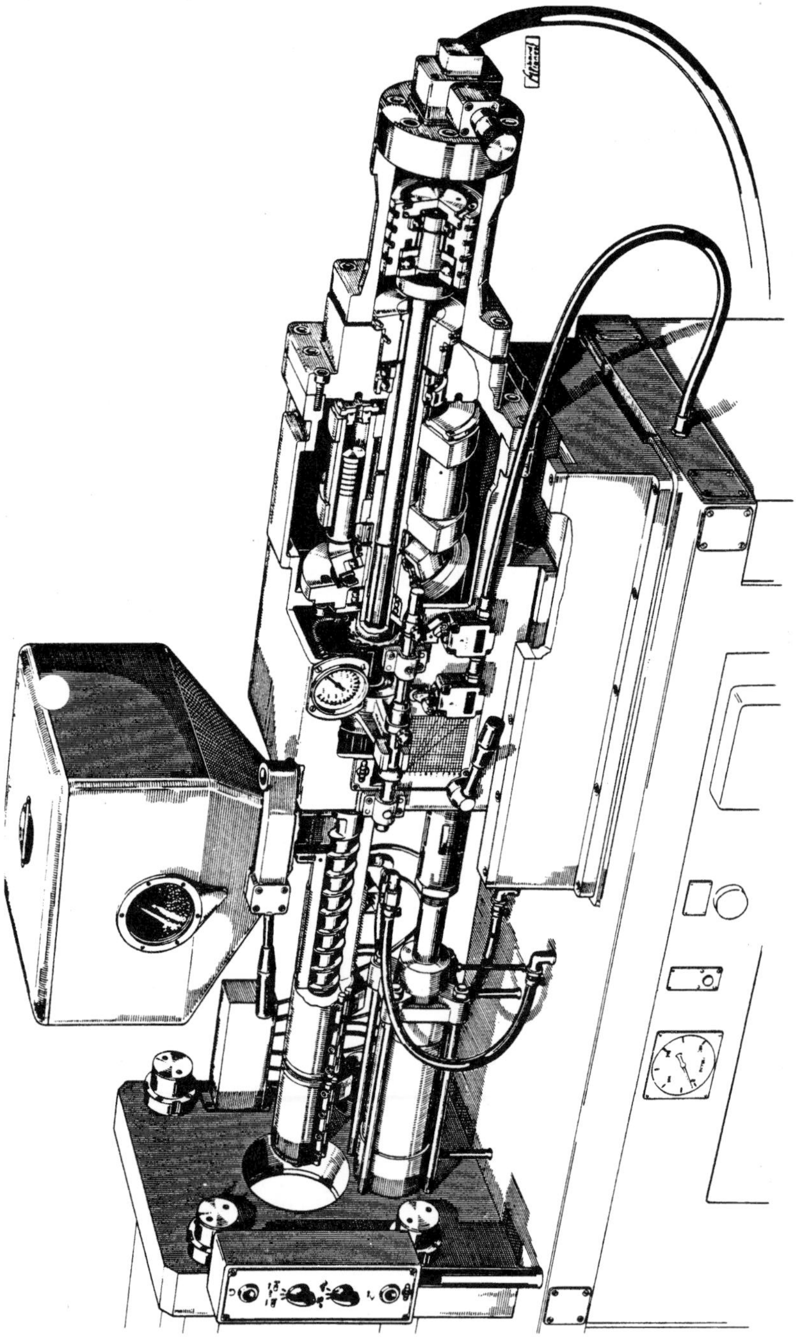

Figure 7.11 Cutout of injection unit or plasticator that melts the plastic compound

The cylinder is a simple heat exchanger. It has heavy steel walls, with highly polished and hardened inner surfaces. Short fiber reinforced plastics require the use of abrasion resistant materials of construction for the barrel and screw; some polymers also require the use of corrosion-resistant materials of construction as well. Examples of materials of construction for reinforced plastics would include nitrided screws and bimetallic alloys for barrels. Hastelloy is often used with fluoropolymers to resist corrosion [3].

It is important to understand that the only direct control input to the melt temperature is obtained through the cylinder heater band settings. The actual melt temperature and its uniformity can vary considerably from these settings, depending on the efficiency of the screw design and the conditions under which the IM machine is operated. Factors that affect the melt temperature include the following:

1. Residence time of the melt in the cylinder;
2. Internal surface heating area of the cylinder (and, where internal heating is provided, the screw) per volume of plastic being heated;
3. Thermal conductivity of both the cylinder and screw, as well as the plastic material;
4. Differential in temperature between the cylinder and the plastic;
5. Wall thickness of the cylinder and of the stationary melt film (on the inner cylinder wall) of the heated plastic;
6. Amount of turbulence in the cylinder; and
7. Effects of shear heating, for example, screw speed (rpm), injection speed, resin viscosity.

Because of their molecular structure, plastics have low thermal conductivity; thus it is difficult to transmit heat through them rapidly. In addition, plastic melts are very viscous, and it is difficult to create any turbulence or mixing action in them without the positive application of some form of mechanical agitation in the screw, for example, rotating the screw faster than the material moves past the flights. The problem is further complicated by limitations on the length of time the plastic may be allowed to remain in the cylinder to minimize degradation.

In choosing or designing a screw, a balance must be maintained between the need to provide adequate time for heating and mixing the material in the cylinder and the need to process maximum throughput for lowest manufacturing costs. It is interesting to note that some reinforcements, for example, carbon fibers, and other additives may increase the heat transfer characteristics of the composite over that of the plastic matrix alone. This faster heat transfer, if properly considered in conjunction with the screw design,

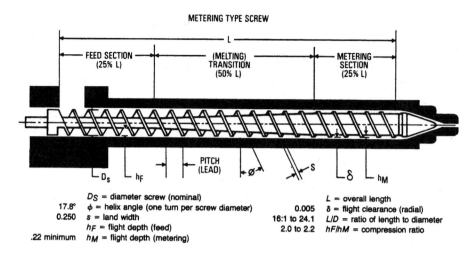

MATERING TYPE SCREW

Figure 7.12 Nomenclature of typical injection molding machine screw

can significantly reduce the time required to heat the melt in the cylinder and cool the melt in the mold.

The primary purpose of using a screw (as compared to a plunger, for example), is take advantage of its ability to mix and knead materials. Theoretically, the motion of the screw should minimize variations in melt temperature. It should also assist in the dispersion of any clumps of reinforcements, colorants, and other additive concentrates, resulting in a more uniform melt being delivered to the mold.

A screw has three basic sections: feed, melting (transition), and metering, as shown in Fig. 7.12. The feed section conveys and heats the material as it enters the cylinder from the hopper. The transition section is where the softened plastic is transformed into a continuous melt. In the metering section, the material is kneaded and sheared to give it a more uniform composition and temperature for delivery to the mold. With extremely heat-sensitive resins, such as PVC, no metering section is used because of the likelihood of causing degradation of the resin. Vented screws are discussed in Section 7.12.

The design of the screw is important for obtaining the desired mixing and melt properties as well as output rate and temperature control. One of the design factors is compression ratio, which is determined by dividing the screw flight depth in the feed zone by that in the metering zone (Fig. 7.12).

Thermoplastics generally require compression ratios of 1.5:1 to 4.5:1; Figure 7.13 provides examples of screw designs for various base resins, using the same length-to-diameter (L/D) ratio for the screw. Thermosets usually

Dimension	Rigid PVC	Impact Polystyrene	Low-density Polyethylene	High-density Polyethylene	Nylon	Cellulose Acet/Butyrate
diameter	4½	4½	4½	4½	4½	4½
total length	90	90	90	90	90	90
feed zone (F)	13½	27	22½	36	67½	0
compression zone	76½	18	45	18	4½	90
metering zone (M)	0	45	22½ (450)	36	18	0
depth in (M)	.200	.140	.125	.155	.125	.125
depth in (F)	.600	.600	.650	.650	.600	

Figure 7.13 Examples of screw designs (1)

require 1:1 compression ratios, affording better heat control and minimizing the possibility of premature curing of the material while still in the cylinder.

There is no "universal" screw for all plastics. Factors such as L/D, length of sections, screw flight land width, clearance between the screw flights and the cylinder wall, etc. all vary with the flow characteristics or rheology of the materials being processed. There are so-called "general purpose" designs, however, that can be used successfully with several different materials, rather than forcing one to inventory a separate (although ideal) screw designed for each product processed. It is particularly important with short fiber reinforced plastics to choose a design that will not cause excessive attrition of the fibers, such as an overly high compression ratio, as this can cause a performance dropoff in the molded part. Consultants and suppliers can recommend optimum screw designs to ensure that part performance is obtained at acceptable processing cost.

7.6　Machine Process Controls

As injection molders become more familiar with their machines and keep up to date on improvements in the machine industry, they will wish to increase the accuracy and sophistication of the controls on their machines. This could enable them to control more closely such features as different injection sequences, inside core mold action, speed of injection, pressure during plastication (back pressure), and mold packing [2].

There are two basic areas of control: the mechanical aspects, such as rpm of the screw, mold opening and closing, etc., and the temperature/rheological aspects, such as cylinder temperature, melt shear, etc. The types of

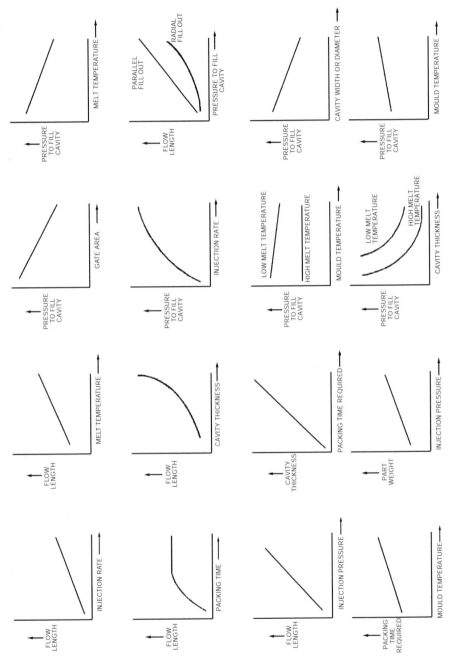

Figure 7.14 Effects of injection molding machine and thermoplastic compounds variables on performance

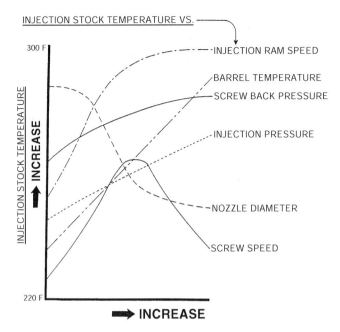

Figure 7.15 Effect of injection molding machine and thermoset compounds variables on performance (melt temperature and stock temperature are used interchangeably)

control used on these aspects can be classed as open loop, closed loop, and adaptive control, which provide increasingly sophisticated methods by which to conduct the IM process and produce parts that meet performance specifications. See Figs. 7.14 and 7.15 for a basic analysis of the effects of specific IM machine and plastic materials variables. The interrelationship of machine and material are shown in Figs. 7.16 and 7.17. These all have to do with the three fundamental properties that must be controlled throughout the process, temperature, pressure, and time.

A typical "start-up procedure" for an IM machine could include the following [2]:

1. Set the temperature controllers on the cylinder and nozzle to recommended values.
2. When the controllers indicate that the settings have been reached, start the machine motors.
3. Close the safety gate and close the press to "lock."
4. Check to see that the resin feed hopper gate is closed and adjust the hydraulic flow control valve to zero setting.

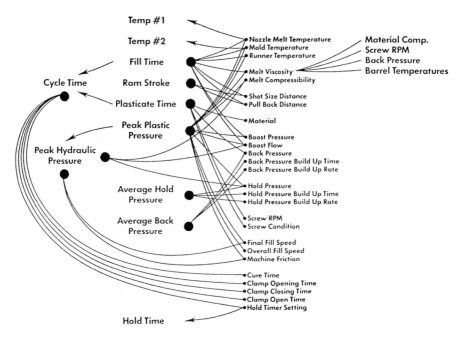

Figure 7.16 Injection molding controls; basically relates to temperature, pressure, and time

5. Turn the plunger switch to the out position and open the hydraulic flow control valve until the screw rotates. If it does not rotate, the material is not yet molten, so turn off the machine and try again in about 10 or 15 min.
6. As the screw rotates, open the feed hopper gate momentarily to allow a small amount of material to feed into the screw. Watch the screw power load and if it reaches or exceeds 100%, reduce screw rpm.
7. Continue opening and closing the feed hopper gate until the material is being ejected more or less steadily and the screw power load appears fairly even.
8. Open the feed hopper gate and run the screw until the ejected melt appears reasonable consistent. Now you are ready to begin molding trial shots.

7.7 Processing Variables versus Molded Part Performance

Short fiber reinforcements and fillers affect molding characteristics significantly and, in turn, are affected markedly by processing parameters [1, 9 to

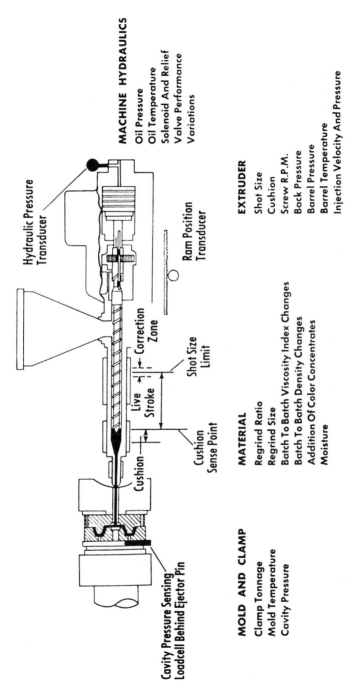

Figure 7.17 Injection process controls (1)

MACHINE HYDRAULICS
Oil Pressure
Oil Temperature
Solenoid And Relief
Valve Performance
Variations

EXTRUDER
Shot Size
Cushion
Screw R.P.M.
Back Pressure
Barrel Pressure
Barrel Temperature
Injection Velocity And Pressure

MATERIAL
Regrind Ratio
Regrind Size
Batch To Batch Viscosity Index Changes
Batch To Batch Density Changes
Addition Of Color Concentrates
Moisture

MOLD AND CLAMP
Clamp Tonnage
Mold Temperature
Cavity Pressure

Hydraulic Pressure Transducer

Ram Position Transducer

Correction Zone

Shot Size Limit

Live Stroke

Cushion Sense Point

Cushion

Cavity Pressure Sensing Loadcell Behind Ejector Pin

12]. In general, fibrous and particulate matter in the polymer matrix increase apparent viscosity and impede melt flow (roughly proportional to concentration). The effect increases dramatically with fiber length and requires higher temperatures and/or pressures to process properly. As reviewed previously, at high reinforcement concentrations, flow is so retarded as to require the use of a ram or plunger, rather than a screw (Fig. 7.6). Combinations of screw and ram can also be used to provide some degree of melting/mixing action prior to flowing into the space in front of the ram/plunger (or "stuffer box") [2, 4].

There are many grades of short fiber reinforced plastics, each tailored to provide a specific set of properties. This allows the design engineer considerable freedom in selecting a material that closely matches part performance requirements. Overall success, however, depends not only on selecting the correct grade, but also on being aware how molding conditions affect mechanical properties, appearance, dimensions, and other aspects of the molded parts.

In some situations, the effects of gating and molding conditions can be substantial. For example, by varying the gating geometry and/or basic molding parameters of temperature, pressure and/or time, different directional properties are obtained. Depending on how the machine is set up and the type of screw used, extensive attrition of the fibers can occur during processing, with attendant reduction of such properties as toughness and strength. In general, low screw speed and low back pressure are desirable to avoid excessive fiber length reduction. Where the gate, melt zones, and cooling channels are located with respect to the mold cavity, can influence localized part properties in the vicinity of the gate and downstream in the cavity.

In short, it is just as important to choose the right molding conditions as it is to choose the right material [5, 6, 12]. Both of the selections are critical to meeting part performance requirements. As in other methods of processing plastics, trade offs are inevitable when using IM. Many variables influence the end results and some variables interact. The most significant molding-related variable in short fiber reinforced plastics is inappropriate or excessive orientation of the fibers. Other potential problems stemming from improper molding include polymer degradation, overpacking of the mold cavity with material (with attendant influence of molded-in stresses), and irregular, excessive, inadequate, or uneven cooling [1, 2, 12].

7.8 Orientation

When the plastic melt is injected into the mold, not only the fiber reinforcements but also the polymer molecules must disengage from each other and orient themselves parallel to the direction of flow to present the least resistance to sliding past each other. Orientation is most pronounced when high-speed flow takes place through narrow gates and thin, confined wall sections, producing high rates of shear and, therefore, high degrees of orientation. After entry into the mold, thicker sections and greater temperature differentials (because there is more distance between the cool mold and the center of the part) permit the stressed polymer melt to relax, reducing the degree of orientation, at least in the center of the part. That portion of the melt that is in contact with the cool mold surface as it flows into the cavity solidifies first, with orientation in the flow direction.

To reduce the degree of orientation found in short fiber reinforced thermoplastics, one can select formulations that also contain glass beads or other particulate fillers. Properties that are strongly directional (due to orientation of the fibers) are termed anisotropic; those that are relatively evenly balanced in all directions are termed isotropic [5, 6, 12]. Orientation increases such properties as strength, stiffness, and toughness, when measured in the direction of flow; correspondingly, they are diminished when measured at right angles to the direction of flow. This phenomenon can be used to advantage by the design engineer where the applied stresses to a part will be unidirectional, by maximizing orientation of the material in the part wall in the direction of the applied stress. When orientation is undesirable, raising the mold temperature and reducing the rate of fill will help. Increasing the gate diameter and the wall thickness of the part will also help, by reducing shear rate. Note, however, that these changes may also result in increased cycle time and possibly greater material usage.

7.9 Weld Lines

During cavity fill, the flow of the melt is obstructed by any cores used to produce holes that are part of the component design. The melt splits as it surrounds the core, then reunites on the other side and continues flowing until the cavity is filled. The rejoining of the split melt forms a "weld line" that has reduced strength and toughness, compared to a homogeneous section. Multiple gating will also cause the formation of weld lines and should be avoided wherever possible. Although the loss of strength and toughness associated with weld lines will vary with the particular base polymer involved, the primary cause is the misalignment of the reinforcing fibers at the face of the meeting flow fronts [2, 9].

For example, 30 weight % glass fiber reinforced polypropylene may show a 66% loss of tensile strength at the weld line, whereas 30% glass fiber reinforced nylon 6/6 may show only a 44% reduction. This factor must be kept in mind during part and tool design. During molding, it is possible to optimize weld line strength by increasing injection hold time. Keep in mind that parts with weld lines have been molded successfully for many years. "Long glass" and other long fiber materials, however, can be particularly difficult to use successfully in parts that unavoidably involve a weld line; it may be preferable to select a short glass compound and design around the problem in such applications.

7.10   Other Variables

With or without reinforcements, excessive heat on the polymer results in degradation (molecular weight reduction, oxidation, etc.), which in turn, reduces properties. In addition to merely setting the temperature controls too high, it is possible to cause degradation through excessive shear, which can be caused by too high a screw compression ratio, too little screw flight to cylinder wall clearance, cracked flights, restrictive check rings and nozzles, undersized runners and gates, or a combination of the foregoing. Also to be avoided are "dead spots" in the polymer flow channels where the melt can accumulate and decompose over time.

Reinforcements and additives can separate from the polymer during flow. When any suspension of solid particles (including fibers) in a liquid (melt) flows, the solid particles tend to concentrate at the front of the flow. While such a concentration is not a frequent problem, different tests can determine if an unacceptable distribution of components exists in a part [1, 2, 5, 6]. Again, part and mold design influence this condition, as will mold temperature, injection rate and pressure, and machine condition.

7.11 Tolerances

Very few IM parts can be held to such extremely close tolerances as less than 2.5 thousandths of a cm or one thousandth of an inch. Tolerances that are used in normal practice go from 5% for 0.05 a cm (0.020 in.) thickness, 1% for 1.27 cm (0.500 in.) thickness, 0.5% for 2.54 cm (1 in.) thickness, or 0.25% for 12.7 (5 in.) cm thickness.

Dimensional accuracy values that can be consistently met depend on such factors as the accuracy of the mold dimensions, the condition and operation of the mold and machine, the type of compound, etc. See Table 7.1 for the parameters that influence part tolerances. Among thermoplastics, amorphous resins are preferred over semicrystalline resins because they typically exhibit lower mold shrinkage. Thermoset materials are generally more suited to meeting tight tolerances than thermoplastics, because they are usually more highly filled and therefore have still lower mold shrinkage [1, 2, 5 to 8]. Of course, thermoset resin selection also plays a role, as the curing step may also affect mold shrinkage.

7.12 Drying Plastic Materials

Plastic materials, particularly thermoplastics, either in virgin forms (such as pellets and granules) or in the form of "regrind" (granulated scrap), are subject to contamination by moisture. When moisture is present in the material during molding, it can cause such defects in the molded part as variable weight, splay marks, brittleness, lower physical properties (as shown in Fig. 7.18), nozzle drool between molding shots, foamy melt, bubbles in parts, poor shot control, and sink marks.

Thermoplastic materials that may absorb moisture and those that are degraded by moisture at molding temperatures must be dried before molding. A drying temperature is selected that will permit the moisture to be driven off without causing the granules to stick to each other, a condition that can cause bridging in the feed throat where the screw receives the material. It is also useful to set the water valve (where available) for cooling the feed throat so that its temperature will not be so low as to cause condensation, or so high as to cause bridging. Attention to the correct setting of the water valve can yield savings in energy (to cool the water and preheat the plastic material). The preferred method of drying plastics is the

Table 7.1 Parameters That Influence Part Tolerance

Part design:	Part configuration (size, shape) relates flow of melt in mold, which in turn affects part performance, including dimensional tolerances.
Material:	Chemical structure, molecular weight, amount and type of additives/fillers, heat history, storage, handling
Mold design:	Number of cavities, layout and size of cavities/runners/gates/ cooling lines/side actions/knockout pins/etc., relate layout to optimized performance of melt and cooling flow patterns to meet part performance requirements. Preengineer design to minimize wear and deformation of mold (use suitable materials of construction), layout cooling lines to meet temperature-to-time cooling rate of plastic material (especially crystalline types).
Machine capability:	Accuracy and repeatability of temperature/time/speed/pressure controls of injection unit, accuracy and repeatability of clamping force, flatness and parallelism of platens, even distribution of clamping force on all tie rods, repeatability of controlling pressure and temperature of hydraulic oil, minimized oil temperature variation, no oil contamination (by the time oil contamination is seen, damage to the hydraulic system is likely to have already occurred), machine properly leveled.
Molding cycle:	Setup the complete molding cycle to repeatedly meet performance requirements at the lowest cost by interrelating material/mold/machine controls as described in the text.

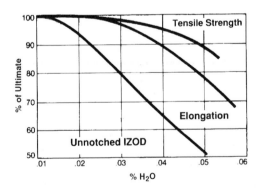

Figure 7.18 Example of the effects of moisture on the mechanical properties of hygroscopic PET plastic during injection molding

dehumidification process whereby moisture is removed and dry air supplied at the specific conditions required for the material being molded.

Plastic materials may be classified into two categories with respect to moisture:

1. Nonhygroscopic. These are polymers where moisture adheres only to the surface of pellets or granules and is not absorbed into the material. Polystyrene (PS), polypropylene (PP), and polyethylene (PE) are examples of nonhygroscopic materials. These polymers may be dried simply by blowing hot air over the material to evaporate the moisture and remove it from the drying unit.

2. Hygroscopic. These polymers absorb moisture into the pellets or granules from their surroundings and form a molecular bond within the material. Common hygroscopic materials include polycarbonate (PC), nylon or polyamide (PA), polyurethane (PUR), polyethylene terephthalate (PET), and acrylonitrile–butadiene–styrene (ABS). These resins can be dried only by removing the moisture from the material using *dehumidified* hot air. Polymers of this type require moisture to be removed before they can be converted into acceptable molded parts. Drying hygroscopic materials cannot be undertaken casually. Simple tray dryers (so-called "pizza ovens") or convection-type hot air dryers, although adequate for nonhygroscopic materials, simply are not capable of removing water to the degree necessary for the proper processing of these materials, particularly during periods of high ambient humidity. If inadequately dried material is heated to the molding temperature, the result can be a chemical reaction between the polymer and water that reduces molecular weight. For example, if "wet" PUR is processed at temperatures in excess of 160 °C (320 °F), its mechanical properties will be seriously degraded. Most PC manufacturers recommend that the moisture content of PC to be molded should not exceed 0.01% by weight (this may vary for specific grades). Since the equilibrium moisture content for PC at average ambient conditions, 73 °F (23 °C) and 50% relative humidity, is 0.18 %, drying is almost always required. PET, as received, may contain 0.05% moisture; to prevent loss of mechanical properties, the resin must be dried so that it contains less than 0.005% moisture before being molded. Although hygroscopic materials are customarily delivered from the manufacturer properly dried and ready for processing in appropriate packaging (such as vacuum-packed, sealed, lined bags or boxes), moisture can be adsorbed from the surrounding air after the packaging is opened. Often this is the result of condensation, such as when material that has been stored in a cold warehouse is brought into a warm (humid) processing area.

Some common equipment used with both hot-air and dehumidifying drying are the air diffuser cone and drying hoppers. As an example, the air diffuser designed to be used in existing loading hoppers consists of a foot-ball-shaped satellite with four legs that permit it to be placed in the center of the hopper. The dry air is brought into the hopper by a flexible hose connected to a hood on top of the hopper, leading to the satellite diffuser.

Nonhygroscopic plastics are usually dried by using a hot-air dryer and plenum hopper or an air diffuser assembly. A hot-air dryer is a relatively simple type of auxiliary equipment that consists of heating elements and an air blower. It can deliver hot air thermostatically controlled up to 149 °C (300 °F) at a typical capacity flow range of 1.7 m³/min (60 to 1,000 ft/min).

Hot-air dryers operate by pulling ambient air into the air-drying filter, through the blower, then across the heating elements. The hot air is then blown to the hopper through a flexible tube. Once the hot air gets to the hopper, it flows among the plastic granules. Hot air performs two functions. The hot air that passes through the plastic evaporates the moisture and moves the resulting water vapor out of the hopper. The hot air also serves to preheat the plastic material. Preheating brings the plastic closer to the molding temperature. When this available heat is used, less heat is needed in the melting process, with a corresponding reduction in energy consumption.

Different factors have to be considered when sizing a hot-air drying system. The first is the plastic material's characteristics. The material will have a specific residence time (length of time the material should be in a hopper dryer), and a specific temperature of the drying air, to ensure thorough drying, without melting or undue softening of the material in the hopper. Another consideration when drying nonhygroscopic plastics is the molding machine production rate or, in simpler terms, the amount of material processed in a 1-h period. If one takes these two factors (residence time and production rate) into consideration, a properly sized plenum hopper can be selected. Then the granules will enter the hopper and slowly work their way down to the bottom of the hopper over the course of 1.5 h (most polymers have a recommended residence time of 1.5 h) and keep up with the set production rate.

Hygroscopic plastics must be dried by using a dehumidifying dryer. These dryers absorb the moisture within the plastic material by using dehumidified, heated air that has a dew point of −40 °C (−40 °F). This is accomplished by using desiccant beads, also called molecular sieves, which are crystalline metal aluminosilicates. A singular feature of these beads is that there is very little dimensional change when water is removed or added. Molecular sieves can dry materials to moisture contents as low as 35 parts per billion (ppb).

Closed loop systems are particularly desirable because they send mois-
ture-laden air through a filter, eliminating any fines that may be present in
the material. The clean, dry air is then reused in the drying process.

There are two additional classifications for drying systems: single-bed
absorption (uses one desiccant bed) and multibed absorption systems (uses
two or more desiccant beds). Air is brought in through a filter on startup and
sent to the desiccant bed to absorb water vapor from the air. Approximately
4.2×10^6 J/kg (1,800 BTU/lb.) is released from the heat of absorption, causing
the air temperature in the typical dryer to rise approximately 11 °C (19 °F).
This warmed air then travels to the heating unit where the air temperature is
raised to the drying temperature required. The heated, dehumidified air is
next circulated through the plastic in the sealed drying hopper. Then the air
is brought out of hopper and recycled back through the drying system, and
the process repeated.

Eventually the beads become saturated with moisture and have to be
regenerated. This is done by blowing air heated to a temperature of 288 °C
(550 °F) through the desiccant beds. The elevated temperature drives the
moisture out of the beds and into the air. This process may vary with different types of dehumidifying dryers.

Multiple desiccant bed absorption is the most efficient method for drying. The commonly used double-bed system utilizes one bed online drying
material, while the other bed is in the regeneration cycle.

Dehumidifying dryers are sized similarly to hot-air drying systems. The
hopper is sized by the production rate multiplied by the residence time. The
dryer is then sized by the corresponding figures from a dryer sizing chart
(supplied by manufacturers). Usually dryers are designed for a flow rate of
0.25 m/sec (50 ft/min). If the flow rate is more than 0.25 m/s, the material will
be blown around in the hopper. Flow rates much lower than this may not
have enough velocity to thoroughly dry the granules.

Another factor to be taken into consideration is the total number of
molding machines that will be running the same material. If more then one
machine will be used, it may be advantageous to set up a central drying
system with one large dryer and a central plenum drying hopper. A central
dehumidifying dryer can also be used with individual high-efficiency plenum
hoppers.

The most effective and efficient predrying system for hygroscopic plastics
is one that incorporates an air dehumidifying system in the materials stor-
age/handing equipment network. Although this type of equipment is initially expensive, it can result in improved production rates and lower reject
levels over a period of time. There are a number of manufacturers and

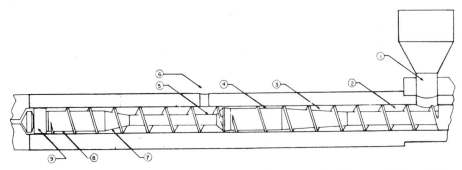

Basic operation of a vented barrel. (1) Wet material enters from a conventional hopper. (2) The pellets are conveyed forward by the screw feed section, and are heated by the barrel and by some frictional heating. Some surface moisture is removed here. (3) The compression or transition section does most of the melting. (4) The first metering section accomplishes final melting and even flows to the vent section. (5) Resin is pumped from the first metering section to a deep vent or devolatilizing section. This vent section is capable of moving quantities well in excess of the material delivered to it by the first metering section. For this reason, the flights in the vent section run partially filled and at zero pressure. It is here that volatile materials such as water vapor, and other nondesirable materials, escape from the melted plastic. The vapor pressure of water at 500°F is 666 psi. These steam pockets escape the melt, and travel spirally around the partially filled channel until they escape out the vent hole in the barrel. (6) Water vapor and other volatiles escape from the vent. (7) The resin is again compressed and pressure is built in the second transition section. (8) The second metering section evens the flow and maintains pressure so that the screw will be retracted by the pressure in front of the non-return valve. (9) A low resistance, sliding ring, non-return valve works in the same manner as it does with a nonvented screw.

Figure 7.19 Schematic of two-stage vented barrel used with an injection molding machine

systems from which to choose. Although, in principle, all systems are designed to accomplish the same thing, the approaches to regeneration of the desiccant beds vary widely. Of interest may be that, in the author's years (decades) of field experience with these systems, it has been observed that breakdowns in performance are very rarely the fault of equipment design, but almost invariably due to the lack of the preventive maintenance that is detailed in the equipment manufacturer's service manuals.

With hygroscopic materials, a more sophisticated approach than a pre-drying system may be the use of a vented barrel molding machine. In some applications that require precision parts meeting very tight performance requirements, *both* predrying *and* molding with a vented barrel may be needed. With a vented barrel (Fig. 7.19), the polymer is devolatilized after it has been melted, and, because the vapor pressure of water at typical melt temperatures is high, devolatilization is usually accomplished rapidly. More-over, at typical melt temperatures, other undesirable nonaqueous volatiles may also be removed by using a vented barrel molding machine. Devolatilization from the melt stream is made possible by the use of a two-stage screw and barrel incorporating a vent port.

The first-stage of the two-stage screw accomplishes the basic plasticating functions of solids feeding and melting. During this first stage process, significant material pressures are generated. Molten plastic leaving the first stage enters a decompression section with a large cross-sectional area such that this channel does not completely fill with melt. As a result, the melt pressure drops to essentially atmospheric pressure, and volatiles (moisture, etc.) are released from the exposed surface of the melt by diffusion. These volatiles are released or escape through the vent opening located in the barrel within the decompression section of the screw. At the end of the second stage, the melt is again compressed to generate the pressure necessary for the melt to flow through a non-return (or other type) valve and provide the force necessary for screw retraction.

The molding operation for a vented barrel IM machine is the same as for a conventional IM machine. However, because the function of melt devolatilization occurs, the shot preparation process involves factors not otherwise encountered, which must be addressed to gain the best performance.

As stated earlier, excessive moisture in plastic materials will reduce or gravely harm the performance of molded parts. Another situation where moisture can be damaging, is the surface of the mold. Molding even non-hygroscopic plastics under high humidity conditions can cause serious problems during production. Normal mold cooling temperatures may produce condensation on the mold surface that can cause appearance defects on molded parts, and over a period of time, corrode some types of molds. Cycle times, scrap rates, operating and maintenance costs, and energy usage upon analysis may point out a previously hidden effect from condensation that affects the entire molding operation all the way from the sales to the bottom line in accounting. Fortunately, this is usually a seasonal or a geographic problem, with which most experienced processors have learned how to cope.

7.13 The Complete Injection Molding Operation

To fabricate all sizes and shapes of parts to meet performance requirements, the molding operation has to fit within the complete operation according to the "FALLO" approach (*Follow ALL Opportunities*), as shown in Fig. 7.20 [1, 2]. Basically, the FALLO approach consists of the following steps:

Relate complete plant operating performance to meet cost-to-performance requirements of molded parts.   by D. V. Rosato

Contents: BASIC PROCESS DESCRIPTION
PLASTICIZING
SCREW DESIGN
DEVELOPING MELT & FLOW
CONTROL
PROCESS CONTROL
INCREASE PRODUCTION THROUGH
FINE TUNING
PURCHASING MATERIALS
CHECKING GOODS RECEIVED

WAREHOUSING
IN-PLANT TRANSPORTATION OF
MATERIALS
IN-PLANT TRANSPORTATION OF
PARTS
PARTS HANDLING EQUIPMENT
COMPOUNDING & COLORING IN-
PLANT
DRYING HYDROSCOPIC PLASTICS
GRANULATING

PRODUCTION CONTROL OF
QUALITY
MACHINE & PLANT SAFETY
ECONOMIC CONTROL OF EQUIPMEN
COSTING
WORK SMARTER-INNOVATE
ANALYZE YOUR FAILURES TO GAIN
MORE SUCCESSES
VALUE ANALYSIS
SUMMARY

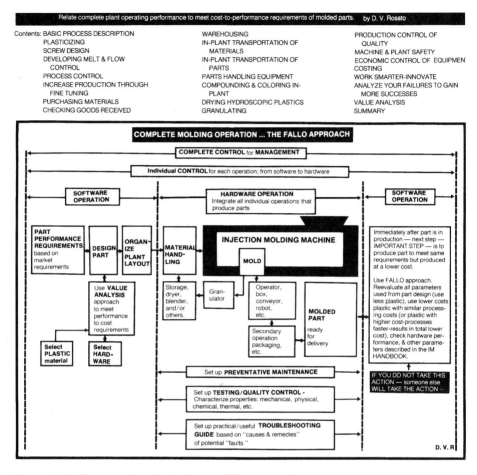

Figure 7.20 The complete injection molding operation

1. Designing a part to meet performance requirements at the lowest cost;
2. Selecting the plastic material that processes into a part that meets performance requirements;
3. Specifying equipment requirements by
   a. designing the mold around the part,
   b. putting the proper capability IM machine around the mold,
   c. selecting auxiliary equipment (material handling, mold temperature controller, granulator, dryer, decorating equipment, etc.); and

    d. Setting up complete controls to produce zero defects (material and part quality testing, IM machine, mold maintenance and trouble-shooting); and

4. Purchasing and inventorying materials and finished parts.

References

1. Rosato, D.V. *Rosato's Plastics Encyclopedia and Dictionary,* (1993) Hanser, Munich
2. Rosato, D.V., Rosato, D.V. *Injection Molding Handbook,* 2nd edit. (1995) Chapman & Hall, New York
3. Dowler, B. "Tool Surface Enhancements: The Extra Edge in Injection Molding," *Inject. Mold.* pp. 79–88, November 1993
4. Lubin, G. *Handbook of Composites* (1982) Van Nostrand Reinhold, New York
5. Dorgham, M.A., Rosato, D.V. "Designing with Plastics and Advanced Composites," *Proceedings of the International Association for Vehicle Design* (1986) Special Publication SP6, Interscience Enterprises, New York
6. Rosato, D.V., et al. *Designing with Plastics and Composites, a Handbook* (1991) Chapman & Hall, New York
7. R. Liebold, "Injection vs. Compression Molding for Large Area GRP Components," *Plast. Machin. & Eng.* pp. 19–24, April 1985
8. Jellinek, K., Bollig, F.J. "The Future for Thermoset Molding Materials," *Eur. Plast.* pp. 25–28, March 1985
9. Cloud, P.J., McDowell, F., Gerakaris, S. "Reinforced Thermoplastics: Understanding Weld-Line Integrity," *Plast. Technol.* August 1976
10. Wuttke, B. "Jute Fibers—an Alternative to Glass Fibers in Reinforcement of Polypropylene," *Kunststoffe,* November 1994
11. Cary, P. "The Asbestos Panic Attack," *U.S. News & World Report,* February 20, 1995
12. Rosato, D.V. "Injection Molding Higher Performance Reinforced Plastics Composites," Society of Plastics Engineers Annual Technical Conference, Indianapolis, IN, May 1996
13. Rosato, D.V. *Designing with Reinforced Composites,* (1997) Hanser, Munich

8 Other Types of Processing

8.1 Compression and Transfer Molding

Compression and transfer molding have been the usual methods employed to fabricate thermoset compounds for more than 100 years. Compression molding is an old, slow process, and not easily automated. Nevertheless, it is a relatively inexpensive way to make a small number of parts because the equipment is simple and tool costs are low. However, as the market opportunities grew for more complex, as well as less expensive parts, thermoset molding machine makers came up with transfer molding as a way to meet this need, beginning in 1926 [1]. Transfer molding also offered improved productivity vs. compression molding because the material feeding step was automated. Since then, transfer molding machines have evolved into units that are very similar to thermoplastic injection molding machines, as described in Chapter 7.

Compression molding is a simple process, consisting of placing the granules of thermoset molding compound into the open lower half of the mold, closing and clamping the mold, heating it to produce crosslinking ("curing") of the compound, opening the mold, and removing the part; see Fig. 8.1. Because the material is placed in an open mold, it is nearly impossible to avoid adding excess (to avoid a partially formed part). When the mold is clamped shut, this excess resin squeezes out of the cavity along the parting line of the mold halves, forming "flash" that must be removed from the part in a secondary operation. It is also possible to make parts from many thermoplastic-based compounds by compression molding. The process is used to best advantage for making a small number of parts that do not have heavy wall sections, and where reinforcing fiber orientation (from flow) is particularly undesirable. Compression molding is much more commercially significant for very long fiber reinforced materials, where injection molding would be impractical because these materials hardly flow at all.

A technique used to speed up the compression molding process is to preheat the molding compound, using an oven or electrical induction heater; this cuts the rest of the molding cycle time. Care must be taken not to overheat, or the compound may begin to flow or even crosslink before it is placed in the mold. This step is more efficient when it is incorporated into the equipment by using a preheating chamber into which the material is placed.

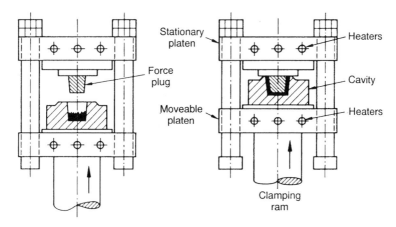

Figure 8.1 Schematic of compression mold and press

After the preheating step, the material is transferred by plunger or screw through a channel to the (closed) mold, hence the use of the term "transfer" molding. The use of a preheating chamber that is directly connected to the mold makes it not only possible, but necessary, to heat the compound to its flow point (but, obviously, not to the point of crosslinking). The introduction of the material into a closed mold, rather than an open one, also eliminates much of the resin "flash" waste. Although more productive than compression molding, transfer molding is also somewhat limited with respect to the compounds that it can efficiently process; the inherent resistance to flow in very long fiber reinforced thermosets makes the transfer step difficult. Figure 8.2 shows a schematic of an early screw transfer machine.

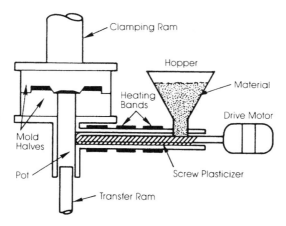

Figure 8.2 Schematic of a screw transfer molding machine

These processes are mentioned because they are still important in thermoset molding, although injection molding is growing rapidly because of the significant cost advantages it offers when part volumes are measured in thousands rather hundreds of units.

8.2   Reaction Injection Molding (RIM)

Reaction injection molding (RIM) is a process mainly used with thermoset polyurethanes; polyureas, polyesters, and even nylon 6 are also molded via this process. Reinforced reaction injection molding (RRIM) is an adaptation of RIM with short fiber reinforcements incorporated in the thermoset matrices mentioned. RIM is a relatively new process, only having come into significant commercial use about 20 years ago. RRIM was developed a few years later. The initial application that brought RIM into commercial significance was flexible polyurethane automobile bumper fascia. RRIM was first used to make PUR automobile fenders and body panels. Polyurethanes and polyureas account for about 95% of materials used in RIM; half of all RIM volume is RRIM [2].

In the case of polyurethanes, the two principal components used to make this thermosetting material, polyol and isocyanate, are liquids held in day tanks where they recirculate through heat exchangers at low pressure. This is to ensure that uniform temperatures are maintained. The reinforcing fibers, usually glass, are mixed into the polyol in the day tank. The polyol is also usually the carrier for the other components, such as an oligomer, catalyst, and additives, such as blowing agents, pigments, etc.

When the molding cycle starts, the two liquid components are fed into metering pistons, which then force the ingredients under high pressure into a mixing head, where they impinge upon each other at high velocity, thus assuring thorough and intimate dispersion. The mixture, now beginning to react, flows under pressure into a mold. Additional mixing is usually provided by incorporating an "aftermixer," or turbulent flow channel, within the mold base but between the snout of the mixing head and the mold cavity itself. Finally, the part is allowed to partially cure, that is, the reaction between polyol and isocyanate has gone sufficiently forward to allow the part to be removed without damage; in a typical cycle, this is only one third of the total, perhaps as little as 30 s. Usually, post-mold curing

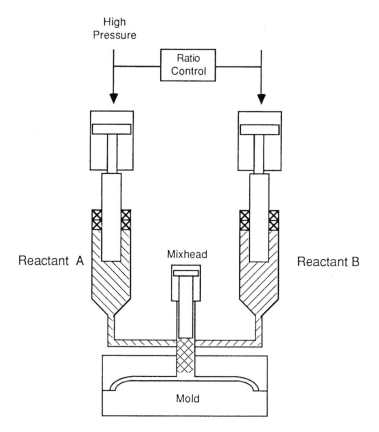

Figure 8.3 Schematic of a typical RIM machine

(crosslinking) will continue for as much as an hour after demolding. Figure 8.3 shows a schematic drawing of a RIM machine.

Milled glass fiber is the most commonly used reinforcement in RRIM, as it will disperse and flow more readily than longer fibers. Nonglass fibers are rarely used in most RRIM applications, largely because of cost considerations. As mentioned earlier, polyurethane is the predominate polymer used in RRIM, both in rigid and elastomer forms. Nylon 6 RRIM is an adaptation from casting technology, utilizing ε-caprolactam with an initiator and a catalyst. Other polymers that can be processed by RRIM include dicyclopentadiene, isocyanurate, acrylamate, epoxy, and phenolic. None of these materials, however, even begin to approach the commercial importance of poly-urethane or polyureas in RRIM.

Advantages of the RRIM process, compared to conventional injection molding include the following:

- Utilizes low clamp-force, less expensive molds.
- Especially suitable for large parts.
- Lower temperatures are required, saving energy costs.
- Accommodates considerable variation in wall thickness.

Although RRIM does have advantages when making parts with varying wall thickness, it is best to avoid unusually thick walls. This is because the liquid components will fill the thicker wall sections first, sometimes leaving air entrapments in the thinner base section, a phenomenon known as "racetracking." If the reason for the thicker wall section is to increase stiffness, there are a number of design techniques that can be used to stiffen side walls without making them thicker than the base, such as curves, chamfering, steps, coring and ribbing, and corrugations. If the part is to be foamed, a space-filling inert insert should be considered for use in unavoidably thick cross-sections [3].

The disadvantages of RRIM compared to conventional injection molding include the following:

- The process is normally not suitable for high fiber loadings, and especially for fibers longer than milled glass, 0.8 mm (0.03125 in).
- Due to the low pressures used, mold surface details can be more difficult to reproduce accurately,
- The low viscosity of the materials can cause flashing, with attendant additional post-finishing costs.
- Handling reactive and often hazardous liquids requires special operator training, equipment, and procedures.
- Mold release is often a problem, requiring the use of external mold release spray.

These limitations can make RRIM more labor intensive than conventional injection molding, so that higher operating costs may offset the savings from lower capital investment. Nevertheless, RRIM has become commercially important because of the advantages cited, rather than the drawbacks.

8.3　PTFE Processing

As mentioned in Chapter 2, the fabrication procedure for PTFE (polytetrafluoroethylene) is significantly different than those used for conventional thermoset or thermoplastic compounds. This is due to the very high molecular weight of the polymer, which reduces its measurable flow to nearly zero. This limitation makes it difficult to produce complex parts from PTFE. Consequently, a substantial number of parts are made by conventional machining techniques, using molded or extruded PTFE rod, sheet, or tubing as the starting material. Production of large numbers of small, relatively simple parts, such as gaskets or bearings, can be and is often automated, using pelletized compounds ("free flow" versus the usual "low flow" pasty mass), which can be fed by a shuttle box into compression molds, rather than fed by hand.

Special care must be taken to avoid contact of PTFE powder with smoking materials, because if the combustion products are inhaled, they can produce a medical condition known as "metal fume fever," which resembles an acute case of influenza lasting 1 or 2 days.

PTFE can be fabricated by simple compression molding without heating, to create a "preform." This preform can be handled but only has a "green" tensile or flexural strength of perhaps 15 to 20% of the strength values it will develop after sintering, when the polymer reaches the gel point and coalesces into a homogeneous structure throughout the part. The last step in the process, cooling, must be conducted at a controlled rate to obtain optimum crystallinity. Compounds that contain high amounts of reinforcement or filler must often be sintered and cooled in the mold under pressure to avoid the formation of cracks and voids in the part. Parts made this way can range from those as small as 1 g or less, to ones weighing several hundred kilograms. Because the processing conditions can strongly affect many of the mechanical and electrical properties of the finished part, it is important to consult the processor, the material supplier, or a consultant, about details [4].

PTFE can also be fabricated by ram extrusion, a process whereby material is fed into the cold end of a long extruder die, which is heated over most of its length. A reciprocating ram compacts the material and forces it into the die, then withdraws before repeating the cycle. Naturally, this process is limited to producing rod, tubing, or profile, which then require secondary fabrication steps, such as cutting and machining, etc. As with molding, ram extrusion processing conditions significantly affect the mechanical and electrical properties of the finished part and the processor, material supplier, or a consultant, should be asked for recommendations. Figure 8.4 shows one type of PTFE ram extruder.

The advantage of ram extrusion over compression molding/sintering is the continuous nature of the process, which can result in reduced production

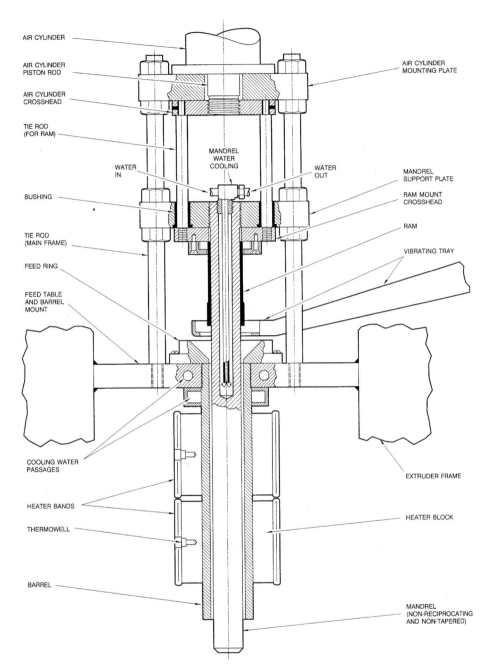

Figure 8.4 Schematic of a vertical ram extruder for making PTFE tubing. (Courtesy of E. I. duPont de Nemours & Co., Inc.)

costs when long runs are involved. The disadvantage is the tendency of some reinforced compounds to "poker chip." The extrusion of reinforced PTFE generally requires high back pressure to reduce void content, accomplished by use of a brake at the end of the extrudate, or a extended mandrel, or a long cooling zone, or a combination of the foregoing. However, too much back pressure will result in circumferential grooves or splits at the charge interfaces, known as "poker chipping" [5].

8.4   Extrusion and Blow Molding

These processing techniques, although widely used for unreinforced thermoplastics, are rarely encountered with short fiber reinforced materials.

Typical single-screw thermoplastic extruders tend to resemble the front of a screw injection molding machine (but without the reciprocating action) as shown in Fig. 8.5. The primary difference is that there is no mold and the material flows from the barrel through a die, either onto a moving belt or through a water bath, where the extrudate is cooled. The extrudate is then either coiled or cut into lengths. The movement of a reinforced material through the extrusion die orients the fibers strongly in the direction of flow, causing such properties as toughness and strength to be quite markedly higher in the longitudinal direction of the extrudate than they are at right angles. Whereas this property differential can run up to 40% in injection molded parts, extruded parts can show an up to 80% variation. In most applications, this difference is unacceptable, hence extrusion is very rarely used for short fiber reinforced plastics, other than the manufacture of semifinished shapes, such as thick rod and sheet, which eventually will be machined into parts. For the reasons mentioned above and in Chapter 5, parts so made will have very different properties than molded parts.

Development of such equipment as counterrotating dies may be successful in overcoming the orientation problem but have not yet reached the commercial stage.

Blow molding consists of extruding a thick tube vertically downward, using a reciprocating screw, immediately clamping the tube between two mold halves, and then introducing pressurized air into the tube while still soft to press it firmly against the mold surfaces. The tube, called a parison or preform, must have sufficient strength in the plasticized state that it does not sag or stretch before it is caught between the mold halves and blown; a

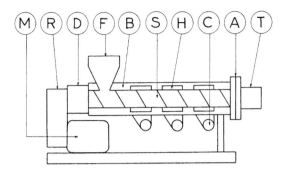

Figure 8.5 Schematic of single screw extruder. (M) Electric motor; (R) Speed reduction device; (D) Thrust bearing; (F) Feed hopper; (S) Screw; (H) Heater; (C) Cooler; (A) Adapter; (T) Tool (die).

schematic illustration of the process is shown in Fig. 8.6. The addition of reinforcing fibers to a thermoplastic usually reduces the "melt strength" of the parison to the degree that it will tear when being formed or blown. Nevertheless, there are a few applications where short fiber reinforced plastics have been successfully formulated and processed by blow molding; the use of a high molecular weight grade of polymer seems to be the solution, although it does increase the cost of the compound slightly. For example, some 30% glass fiber reinforced nylon 66 automobile radiator overflow tanks have been made by blow molding.

The economics of blow molding a unitary hollow object that requires some post-mold finishing (removal of the excess parison where it has been

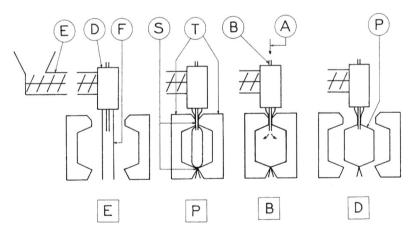

Figure 8.6 Schematic of extrusion blow molding process. [E] Extrusion; [P] Pinching; [B] Blowing; [D] Demolding; (E) Extruder; (D) Die; (F) Preform; (S) Squeezed area; (T) Tool; (B) Blow pin; (A) Air pressure; (P) Product.

clamped at either end of the mold) versus injection molding two halves that are subsequently joined together must be calculated to determine which technique will cost less overall. Usually, blow molds are less expensive than injection molds, but blow molding cycle times are normally longer than injection molding cycle times. Of course, another possibility for making hollow parts is the lost-core injection molding process, but this requires a significant investment in equipment to prepare and recover the cores. Blow molding may be worth evaluating where a modest number of relatively large, hollow parts are required.

The process described above is also called extrusion blow molding, to differentiate it from injection blow molding, a variation whereby the parison is injection molded in one mold, then blown in another mold. However, injection blow molding is typically used to make small bottles and jars; the author is not aware that this process has ever been used commercially with short fiber reinforced plastics.

8.5   Secondary Processing

Secondary processing, also called finishing, includes machining, joining, and decorating parts that have already been manufactured by primary methods, such as molding. The presence of fiber reinforcements in the article to be finished can influence the techniques to be used and the results obtained thereby. Finishing may be as simple as removing unwanted bits of flash from the piece along the mold parting lines, as well as sprues and gates where they are directly attached to the piece. Also, it may be less expensive or technically preferable to drill holes or place inserts as a post-molding step, rather than incorporate holes or inserts in the molding process itself, for example, to avoid weld lines.

The design of cutting tools and their materials of construction needs to be specific to the material to be machined; what is acceptable for drilling carbon steel is unlikely to be acceptable for drilling glass fiber reinforced epoxy. The machining of plastics has grown in use and in sophistication to the point that it has even lead to the publication of a specific journal, *Plastics Machining and Fabricating.* In general, fiber reinforced composites require more care to machine than neat resins because the fibers tend to tear out from the surface, leaving a ragged edge. Slow speeds, generous cooling, and hardened steel tools are typical of the requirements for machining these materials. Manufacturers of stock shapes are usually good sources of information for specific details on how to best machine plastics.

Sometimes it may be less costly to assemble some components rather than to mold them as one, by using snap fits, interference fits, welding, or cementing. The incorporation of metal inserts as a post-molding step, rather than during the molding process, can improve molding machine productivity and permits greater flexibility in choosing the type of insert to be used, for example, expansion, broaching, etc.

Joining thermoplastic parts by welding is a well-established technique, applying heat to the interface so as to soften the polymer on both sides and allow it to flow together. Several methods have been developed to generate this heat, including vibration and ultrasonics, heated plates or tools, and electrical induction. Perhaps the simplest is ultrasonic welding. Short fiber reinforced materials present a problem because, not only is it impossible to avoid a loss of strength and toughness in the joint, compared to that of a molded part, but it may not be possible to retain any higher degree of these properties at the weld than is present in the matrix polymer alone, if the reinforcing fibers are pushed away from the joint as it is formed, or are oriented parallel to the joining line, or undergo severe attrition in length. Nevertheless, studies have shown that if conditions are optimized, such as ensuring that the weld zone is sufficiently wide, joint strengths of up to 50% of the compound's tensile strength can be retained (nearly 20% greater than the unreinforced matrix resin's tensile strength) [6].

In Section 8.4, it was mentioned that hollow parts can be made by joining two molded halves; this technique may be more cost effective than blow molding or lost-core injection molding a single part, depending on design specifics and the volumes of parts to be produced. For example, automobile air intake manifolds made of 30% glass fiber reinforced nylon 66 are mainly produced by lost-core injection molding, but the feasibility of making them from 30% glass reinforced nylon 6 in vibration-welded halves has also been demonstrated. It is interesting to note that blow molding to date has not been considered a feasible alternative process because variations in part wall uniformity have been thought to exceed the very demanding specifications for this application.

Joining by heat alone is not feasible for thermoset-based materials, and adhesives offer the most practical solution. The specific adhesive or cement to be used will, of course, depend on the specific polymer matrix of the composite (the presence of a reinforcement does not usually affect the choice), and the suppliers of both should be consulted about their recommendations.

Post-mold decorating of parts may occasionally be necessary, and can be accomplished by printing, hot stamping, labeling, metallizing, or painting. If these techniques are to be used, then the processor should set molding conditions so as to produce parts with a resin-rich, smooth surface, for optimum adhesion and appearance. This can be accomplished by using a warm mold (which may increase the molding cycle time). Otherwise, most materials exhibit a molded surface that appears frosted and slightly bumpy. Some compounds, particularly those based on nylon 6, usually have a naturally glossy surface under normal molding procedures. Beyond these considerations, decorating short fiber reinforced plastics is very much like decorating their unreinforced versions.

References

1. Wright, R.E. *Injection/Transfer Molding of Thermosetting Plastics* (1995) Hanser, Munich
2. Macosko, C.W. *Fundamentals of Reaction Injection Molding* (1988) Hanser, Munich
3. Oertel, G. *Polyurethane Handbook* (1993) Hanser, Munich
4. Anon. *Compression Molding of **Teflon**® PTFE,* E. I. du Pont de Nemours & Co., Inc., Wilmington, DE, May 1995
5. Anon. ***Teflon**® PTFE Ram Extrusion Processing Guide,* E. I. du Pont de Nemours & Co., Inc., Wilmington, DE, June 1991
6. Kagan, V.A., Lui, S.-C., Patry, J. "Optimizing the Vibration Welding of Glass-Reinforced Nylon Joints," *Plast. Eng.* Sept. 1996

Appendix A: Bibliography

In addition to the listings at the end of the chapters, the following publications have been selected as representative of reference works that have a practical bent for the engineer. The directories provide the most up-to-date listings of products, suppliers, and processors. The computer software listing is representative of just a few of the more widely used data base, design, and engineering programs meant to be run on personal computers rather than mainframes. The reader is cautioned that this is a rapidly evolving field and the programs named may have been replaced or outmoded since this list was assembled.

1 Materials

Titow, W.V., Lanham, B.J. *Reinforced Thermoplastics* (1975) Applied Science, London
Folkes, M.J. *Short Fibre Reinforced Thermoplastics* (1982) Research Studios Press, Letchworth, Herts., England
Clegg, D.W., Collyer, A.A. *Mechanical Properties of Reinforced Thermoplastics* (1986) Elsevier Applied Science, Barking, Essex, England
Charrier, J.-M. *Polymeric Materials and Processing* (1991) Hanser, Munich
Engineered Materials Handbook Desk Edition (1995) ASM International, Materials Park, OH

2 Design

Powell, P.C. *Engineering with Polymers* (1983) Chapman & Hall, London
Levy, S., DuBois, J.H. *Plastics Design Engineering Handbook* Second Edition (1984) Chapman & Hall, New York
Ehrenstein, G.W., Erhard, G. *Designing with Plastics: A Report on the State of the Art* (1984) Hanser, Munich
Malloy, R.A. *Plastic Part Design for Injection Molding* (1994) Hanser, Munich

3 Processing

Rosato, D.V., Rosato, D.V. *Injection Molding Handbook* Second Edition (1995) Chapman & Hall, New York
Macosko, C.W. *RIM Fundamentals of Reaction Injection Molding* (1988) Hanser, Munich
Wright, R.E. *Injection/Transfer Molding of Thermosetting Plastics* (1995) Hanser, Munich

4　　Directories

Modern Plastics Encyclopedia McGraw-Hill, New York, published annually
Plastics Technology Manufacturing Handbook & Buyers' Guide Bill Communications, Inc.,
New York, published annually
Plastics Recognized Component Directory Underwriters Laboratories, Northbrook, IL, pub-
lished annually in April and supplemented in October

5　　History

R.F. Jones, "U.S. Independent Compounding—Past, Present, Future," *Plastics Engineering*
May 1996

6　　Computer Software

CenBASE/Materials, offline database for thermoplastics, thermosets, elastomers, and composites; updated quarterly. Infodex, Inc., 12842 Valley View St., Garden Grove, CA 92645

PLASPEC, online database for thermoplastics and thermosets. D&S Data Resources, Inc.,
218 E. Bridge St., Morrisville, PA 19067

IDES offline database for thermoplastics, thermosets, elastomers and composites, updated
quarterly; also cost estimator. Integrated Design Engineering Systems, Inc., 207 Grand,
Laramie, WY, 82070

Rover Electronic DataBook— Chemical and Environmental Resistance of Plastics and Rub-
bers (CD-ROM version of Plastics Design Library's *Chemical Resistance Handbooks*), William
Andrew, Inc., 13 Eaton Avenue, Norwich, NY 13815

CAMPUS (Computer Aided Material Preselection by Uniform Standards) data merge and
search software:
　MCBase, The Madison Group, 565 Science Drive, Madison, WI 53711
　Polymerge™, Modern Plastics International, P.O. Box 605, Hightstown, NJ 08520

CAMPUS harmonized material properties databases from
　AlliedSignal, Inc., Morristown, NJ
　BASF Corp., Mount Olive, NJ, and BASF AG, Ludwigshaven, Germany
　Bayer Corp., Pittsburgh, PA, and Bayer AG, Leverkusen, Germany
　CYRO Industries, Rockaway, NJ
　Dow Chemical Co., Midland, MI
　DuPont Engineering Polymers, Wilmington, DE
　Eastman Chemical Co., Kingsport, TN
　Elf Atochem N. America, Philadelphia, PA, and Elf Atochem S.A., Paris, France
　Ems-American Grilon Inc., Sumter, SC, and Ems-Chemie AG, Domat, Switzerland
　EniChem America, Houston, TX, and EniChem Spa, Milano, Italy

General Electric Plastics, Pittsfield, MA
Hoechst Celanese Corp., Summit, NJ
Hüls America, Inc., Piscataway, NJ, and Hüls AG, Marl Germany
Nyltech N. America, Manchester, NH, and Nyltech France, St. Fons, France
Ticona (Hoechst Group), Summit, NJ, and Kelsterbach, Germany

Proprietary materials property databases from
General Electric Co., GE Plastics, Pittsfield, MA
RTP Co., Winona, MN
Thermofil, Inc., Brighton, MI

Internet home page addresses
AlliedSignal   http://www.alliedsignal.com
BASF Plastics Materials   http://www.basf.com.plastics
Dow Chemical   http://www.dow.com
Eastman Chemical   http://www.eastman.com
GE Plastics   http://www.ge.com.gep/homepage
LNP Engineering Plastics   http://www.lnp.com
PLASPEC (see above)   http:www.plaspec.com

AutoCAD® series for designing, modeling, and drafting, Autodesk, Inc., 111 McInnis Pkwy., San Rafael, CA 94903

CADKEY series for 3-D modeling, design, analysis, and drafting. CADKey, Inc., 4 Griffin Rd. North, Windsor, CT 06095

C-MOLD series for designing, modeling, and numerous elements of molding analysis, AC Technology North America, Inc., 50 Moore Ave., Waldwick, NJ 07463

MARC series for designing, modeling, non-linear finite element analysis, Marc Analysis Research Corp., 260 Sheridan Ave., Suite 309, Palo Alto, CA 94306

MOLDFLOW series for flow, cooling, warpage, shrinkage and gas penetration analysis, many specific to short fiber reinforced plastics.Moldflow, Inc., 2 Corporate Drive, Suite 232, Shelton, CT 06484

Design for Assembly series for the evaluation of part designs and assembly systems; Design for Manufacturing series for defining product features by cost-efficiency of materials and related manufacturing processes. Boothroyd Dewhurst, Inc., 138 Main St., Wakefield, RI 02879

Appendix B: Short Fiber Reinforced Plastics Suppliers

The following list of commercial suppliers and their product lines is reasonably complete, but is not exhaustive, because changes are constantly taking place. Criteria for selection include polymer production facilities (or at least U.S. compounding), breath and depth of a proprietary prime product line, product development, and international distribution. Many of these suppliers either offer computer diskettes containing databases describing their product line and corresponding properties, or offer similar information via their home page on the Internet.

Reinforced thermoplastics suppliers	Nylon	Styrenics (including PPE blends)	PP/HDPE	Polyesters (including TPE)	PC	POM	Others (see footnote)
Adell Plastics, Inc., Baltimore, MD	X		X	X		X	
Albis Corp., Rosenberg, TX	X		X				
Albis Plastic GmbH, Hamburg, Germany							
AlliedSignal, Inc., Engineering Plastics Division, Morristown, NJ	X			X			
Amoco Polymers Alpharetta, GA			X				X-2,3,6
Asahi Chemical Industry Co., Ltd. Tokyo, Japan	X	X				X	
BASF Corp., Mount Olive, NJ	X	X		X		X	X-2, 4
BASF AG., Ludwigshafen, Germany	X						
Bayer Corp., Polymers Division Pittsburgh, PA	X	X		X	X		
Bayer AG, Leverkusen, Germany							
Bay Resins, Inc., (subsidiary of Clariant Corp.), Millington, MD	X	X		X	X	X	
Borealis A/S, Lyngby, Denmark			X				
Chem Polymers, Division of T&N Industries, Inc., Ft. Myers, FL	X						
Chisso Corp., Tokyo, Japan			X				
ComAlloy International (subsidiary of A. Schulman, Inc.), Nashville, TN	X	X	X	X	X		X-2,5,7
Custom Compounding (division of Dyneon LLC), Aston, PA							X-1
Custom Resins Group Henderson, KY	X						

Company						
Daicel Chemical Industries, Ltd. Osaka, Japan		X				
Diamond Polymers, Inc., Akron, OH		X				
Dow Chemical Co., Midland, MI	X	X				X-4
DSM Engineering Plastics Evansville, IN	X		X	X	X	X-2,4,5,6
E.I. duPont de Nemours & Co., Inc. Polymer Products Department Wilmington, DE	X	X	X	X		X-1,3
Eastman Chemical Co., Performance Plastics Division, Kingsport, TN			X			X-4
Elf Atochem N. America, Technical Polymers, Philadelphia, PA / Elf Atochem SA, Paris, France	X					X-1
Ems-American Grilon, Inc. Sumter, SC / Ems-Chemie AG, Domat, Switzerland	X					
Enichem Spa, Milano, Italy	X	X	X	X	X	
Ferro Corp. Filled and Reinforced Plastics Division Evansville, IN	X	X	X	X	X	
Foster Corp., Dayville, CT, USA	X	X	X	X	X	
General Electric Co., GE Plastics Pittsfield, MA	X	X	X	X		X-5,6
Geon Co., Cleveland, OH						X-7
Hanna Engineered Materials (formerly CTI, Texapol), Norcross, GA	X	X	X	X	X	X-2,5,6
Hoechst Celanese Corp., Technical Plastics Division, Chatham, NJ / Hoechst AG, Frankfurt/M, Germany	X	X	X	X		X-1,3,5,6

Reinforced thermoplastics suppliers	Nylon	Styrenics (including PPE blends)	PP/HDPE	Polyesters (including TPE)	PC	POM	Others (see footnote)
LATI USA Inc., Mt. Pleasant, SC LATI SPA, Vedano Olano, Italy	X	X	X	X	X	X	X-2,5,6
LNP Engineering Plastics, Inc. (subsidiary of Kawasaki Steel Corp.) Exton, PA	X	X	X	X	X	X	X-1,2,3,4,5,6,7
Marval Industries, Inc., Mamaroneck, NY, USA							
Mitsubishi Engineering Plastics Corp. Tokyo, Japan	X	X		X	X	X	X-3
Mitsubishi Rayon Co., Ltd. Tokyo, Japan		X	X	X			
Modified Plastics, Inc. Santa Ana, CA	X	X	X	X	X	X	X-2,4,5,6
Montell Advanced Materials Lansing, MI			X				X-4
Multibase Inc., Copley, OH			X				
Nan Ya Plastics Corp. Taipei, Taiwan	X		X	X	X		
Nova Chemicals (div. Of Nova Corp.) Leominster, MA		X					
Nyltech N. America, Manchester, NH Nyltech S.A., St. Fons, France Nyltech SRL, Milan, Italy	X						
Phillips Chemical Co., Engineering Plastics, Bartlesville, OK							X-5
The Plastics Group, (formerly Polyfil, Inc.), Woonsocket, RI	X	X	X				

Company							Others
PMC (Division of Clariant Corp.) Milford, DE	X	X	X	X	X	X	
Polycom Huntsman Inc. Washington, PA	X			X		X	
Polymer Composites, Inc. (subsidiary of Ticona) Winona, MN	X	X	X	X	X	X	X-3,4,5
Polymer Resources, Ltd. Farmington, CT	X	X	X	X	X	X	
PTFE Compounds, Inc. New Castle, DE					X	X	X-1
Rhodia S.A., St.-Fons, France	X		X		X		
RTP Corp., Winona, MN	X	X	X	X	X	X	X-1,2,3,4,5,6
A. Schulman, Inc., Akron, OH	X		X				
Solutia, Inc. (formerly Monsanto Co.), St. Louis, MO	X		X				
Thermofil, Inc., (subsidiary of Nippon Steel Corp.), Brighton, MI	X	X	X	X	X	X	X-2,4,5,6
Teijin Chemicals Ltd., Tokyo, Japan				X			
Ticona (Hoechst Group) Summit, NJ, USA Kelsterbach, Germany	X		X	X	X	X	X-1,3,5,6
TP Composites, Inc., Aston, PA	X	X	X	X	X	X	
Ube Industries Ltd., Tokyo, Japan	X	X					
Victrex USA, Inc., West Chester, PA Victrex plc, Blackpool, UK					X		X-6
Wellman, Inc., Plastics Division Johnsonville, SC	X		X				

Others: 1-Fluoropolymers; 2-PSU/PES; 3-LCP; 4-TPU; 5-PPS; 6-PAR/PAI/PEI/PEK/PEEK; 7-PVC

Reinforced thermoset suppliers	Alkyds	Allyl (DAP)	Epoxy	Amino	Phenolic	Polyester	Polyimide/ silicone	RRIM Systems
Bayer Corp., Polymers Division, Pittsburgh, PA Bayer AG, Leverkusen, Germany								X-1,2
Bulk Molding Compounds, Inc., St. Charles, IL						X		
Cytec Industries, Inc. West Paterson, NJ						X		
Ciba-Geigy Corp., Polymers Division, Hawthorne, NY Ciba-Geigy Ltd., Basel, CH							X	
Cosmic Plastics, Inc. San Fernando, CA		X	X					
Cook Composites & Polymers, Inc. Kansas City, MO								X-3
Dow Chemical Co. Midland. MI								X-4
DSM RIM Nylon, Inc. Brook Park, OH								X-4
Fiberite Inc., Winona, MN			X	X	X			
ICI Polyurethanes West Deptford, NJ							X	X-1
Glastic Company Cleveland, OH						X		
Haysite Reinforced Plastics Erie, PA						X		
Industrial Dielectrics, Inc. Noblesville, IN						X		

Supplier	RRIM 1-PUR	RRIM 2-Urea	RRIM 3-Polyester	RRIM 4-Nylon 6
Occidental Chemical Corp. Durez Division, Dallas, TX	X		X	
Perstorp AB Perstorp, Sweden			X	X
Plaslok Corp., Buffalo, NY			X	
Plastics Engineering Co. (Plenco) Sheboygan, WI			X	X
Polyply Composites, Inc. Grand Haven, MI	X			X
Premix, Inc., Kingsville, OH	X			X
Resinoid Engineering Corp. Materials Division, Skokie, IL			X	
Rogers Corp. Manchester, CT	X	X	X	X
Rostone Corp., Lafayette, IN	X			X

RRIM: 1-PUR; 2-Urea; 3-Polyester; 4-Nylon 6

Appendix C: Applications of Short Fiber Reinforced Plastics

A   Automotive

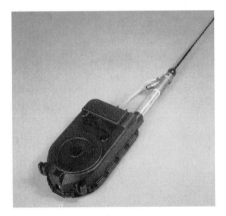

Figure C.A1 Housing for power-operated car radio antenna—25% glass fiber reinforced flame retardant nylon 66. Requirements: high rigidity and toughness over a temperature range of −40 to 110 °C, good dielectric properties; replaces insulated metal housing. (Supplied by BASF)

Figure C.A2 Disk brake caliper piston—50% glass fiber reinforced phenolic. Requirements: very high rigidity and dimensional stability over a similar temperature range as in A.1, high resistance to oil, grease, and moisture. (Supplied by Fiberite)

Figure C.A3 Sports car seat back and pan—50% long glass fiber reinforced nylon 66. Requirements: very high rigidity under load and over a temperature range similar to that in A.1; replaces metal assembly that was twice as heavy. (Supplied by LNP Engineering Plastics)

B   Electrical/Electronic

Figure C.B1 Electronic chip carrier sockets and connectors—40% glass fiber reinforced polyphenylene sulfide. Requirements: high toughness and flexural strength at ambient temperatures, high rigidity and dimensional stability at operating temperatures of up to 150 °C, excellent dielectric properties. (Supplied by Ticona)

Figure C.B2 Telecommunication connectors—40% glass fiber reinforced phenolic. Requirements: High rigidity, dimensional stability, and dielectric properties at operating temperatures. (Supplied by Fiberite)

Figure C.B3 Computer printer chassis—40% glass fiber reinforced and mica filled PPE/PS blend. Requirements: high structural rigidity and dimensional stability, low shrinkage and warpage, flame retardancy, high toughness. Replaces metal chassis and associated hardware (part consolidation). (Supplied by General Electric Plastics)

Figure C.B4 Laptop computer "butterfly" keyboard—30% carbon fiber reinforced polycarbonate. Requirements: high stiffness and toughness, flame retardancy, high flow and high knit line strength when molded, EMI/RFI shielding. Original part design. (Supplied by General Electric Plastics)

Figure C.B5 Airframe electrical interconnect—40% carbon (PAN) fiber reinforced PEEK. Requirements: high stiffness at operating temperatures and potential emergency conditions of more than 285 °C, flame retardancy and low smoke generation, ability to be surface metallized for shell conductivity and EMI/RFI shielding. Replaces metal. (Supplied by RTP)

Figure C.B6 Oscilloscope ring bezel—20% glass fiber reinforced flame retardant polycarbonate. Requirements: high stiffness and toughness at ambient temperatures, flame retardancy, colorability, high flow during molding. Replaces metal. (Supplied by RTP)

C   Consumer

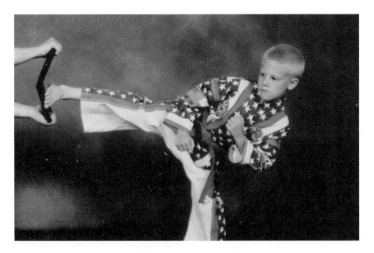

Figure C.C1 Karate kick board—33% glass fiber reinforced nylon 66. Requirements: rigidity and strength similar to wood, but more uniform, must be reusable. Replaces single-use wood board. (Supplied by DuPont)

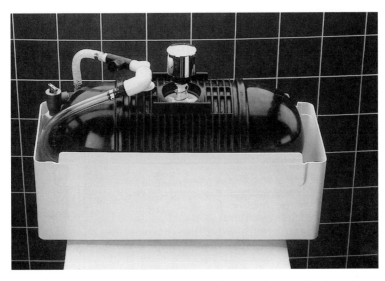

Figure C.C2 Toilet tank—40% glass fiber reinforced thermoplastic polyester (PBT). Requirements: resistance to water absorption or attack at ambient temperature and under 35 psi pressure, excellent creep resistance and rigidity, good surface finish. Replaces metal. (Supplied by Ticona)

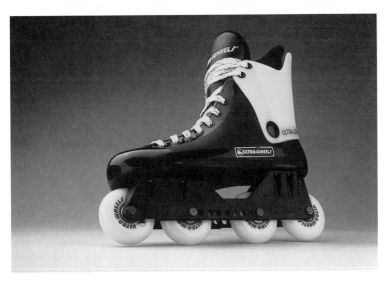

Figure C.C3 In-line roller skate frame—40% long glass fiber reinforced nylon 66. Requirements: high rigidity and toughness. Replaces metal that weighed more than 2.5 times as much. (Supplied by Ticona)

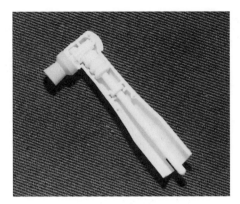

Figure C.C4 Dental equipment bevel gears—30% glass fiber reinforced, PTFE lubricated nylon 66. Requirements: high stiffness at operating (ambient) temperatures, lubricity, high flow during molding. Replaces metal. (Supplied by RTP)

Figure C.C5 Outboard motor propeller—40% glass fiber reinforced thermoplastic polyurethane. Requirements: high toughness and stiffness in an aqueous environment at ambient temperatures. Replaces stainless steel. (Supplied by RTP)

D   Industrial

Figure C.D1 Gear pump impeller—40% glass fiber reinforced epoxy. Requirements: resistance to a wide variety of chemical reagents at temperatures ranging from ambient up to 150 °C, high dimensional stability and rigidity under the same conditions. (Supplied by Fiberite)

Figure C.D2 Cooling fan hub and blades—50% long glass fiber reinforced, impact modified nylon 6. Requirements: high dimensional stability and toughness over an operating temperature range from −50 °C to ambient. Replaced glass fiber/thermoset polyester layup. (Supplied by LNP Engineering Plastics)

Index

Roger F. Jones received a BS with Honors in Chemistry from Haverford College in 1952 and then entered a career in the plastics industry that included technical and management positions in manufacturing, research & development, and marketing, with E.I. duPont de Nemours & Co., Inc. (nylon intermediates), Atlantic Refining Co. (paraffin wax/polyethylene coatings), and Avisun Corp. (polypropylene). During this period, he also served on active duty for three years as an officer in the United States Navy at the end of the Korean War. In 1967, he joined LNP Corporation (short fiber reinforced thermoplastic compounds) and moved up progressively to president; after LNP was acquired by Beatrice Foods Co., he was additionally named a group executive in Beatrice's Chemicals Division. In 1981, he became chairman and president of Inolex Chemical Company (plasticizers and urethane polyols). In 1984, he was appointed as managing director, BASF Corporation Engineering Plastics, and in 1989, he founded Franklin Polymers, Inc., Broomall, Pennsylvania, an engineering plastics distribution and marketing/management consulting firm; he serves as its president.

Mr. Jones is a widely published authority on plastics, particularly short fiber reinforced thermoplastic compounds, although this is his first book. Both in the U.S. and overseas, he is the author of over 75 articles and papers, and inventor of record for 20 patents. His honors include the Honor Scroll of the American Institute of Chemists, Sigma Xi, and election as a Fellow of the Society of Plastics Engineers. He is also a member of the American Chemical Society, Society of the Chemical Industry (American Branch), Fellow of the American Institute of Chemists (past National Secretary, Board Vice Chairman, and Pennsylvania Institute President). He has been a guest lecturer at the Universities of Delaware, Wisconsin, and Toronto, The Packaging Institute, and the Plastics Institutes of England and Australia.

Mitchell R. Jones is a Product Manager at Instron Corporation, Canton, Massachusetts. He previously was in composites research at PPG Industries, Inc. Mr. Jones earned a BSME degree with emphasis on composite materials science from the University of Delaware. He is a member of the American Society for Testing and Materials and has published on the precision and accuracy of laboratory physical testing instruments.

Donald V. Rosato is program director of Plastics FALLO, Waban, Massachusetts. He has previously been market development manager for medical products at Borg Warner Chemicals, Inc., Parketsburg, West Virginia. He holds a PhD (marketing) from the University of California, an MBA from Northeastern University, an MS (plastics and chemical engineering) from the University of Lowell, and a BS (chemistry) from Boston College. He is a member of the Society of Plastics Engineers, American Chemical Society and American Institute of Chemical Engineers, and is widely published on plastics technical and marketing subjects.